There Is No Energy Problem

Coleman Raphael

authorHOUSE®

AuthorHouse™
1663 Liberty Drive
Bloomington, IN 47403
www.authorhouse.com
Phone: 1-800-839-8640

First published by AuthorHouse 9/7/2011

ISBN: 978-1-4567-4966-8 (dj)
ISBN: 978-1-4567-4967-5 (e)
ISBN: 978-1-4567-4968-2 (sc)

Library of Congress Control Number: 2011903540

Printed in the United States of America

This book is dedicated to my daughter, Hollis, and my wife, Sylvia, whose comments and recommendations enabled me to employ logic and common sense in converting analysis of existing data to future projections and recommendations. I am also grateful to the neighbors and friends, such as Carolyn and John McHale, who made helpful suggestions concerning style and content. I hope that governmental, scientific, and economic analysts will be influenced by the arguments presented herein and will support a national energy program that is in the best interests of today's society.

Table of Contents

Introduction and Summary

The purpose of this book is to demonstrate how our forward-looking and progressive nation can be free of dependence on limited energy sources, such as foreign oil. Oil was discovered by accident in the late nineteenth century, and at the current rate of use, it will be gone in less than fifty years. It is time for us to move on and begin to harness the rays of the sun—an inexhaustible supply. We must set this as a national goal and act now. This book defines energy and describes its current uses and sources. After describing and analyzing the pros and cons for each of these sources, we will ultimately recognize how we can economically and efficiently use the sun's energy to supply all of our current and future needs.

We live in a society and an era where the term "energy" plays a major role. We are impressed with the energy displayed by a tap-dancing acrobat, or the "lack of energy" in an individual who just wants to lie on a couch, or the massive release of energy when a bomb explodes, or the energy contained in a speeding vehicle. More seriously, society has become concerned about the availability of energy. Is oil a major source of energy? What will we do when our supply of oil is depleted? What role does the sun play? What other concerns should we have about the world's energy supply?

This book shows that the earth is in a condition of energy equilibrium, where the total energy coming in (from the sun) is equal to the energy radiating away, so that the average temperature at the earth's surface remains constant (at about 59 degrees Fahrenheit). This "energy balance" for the earth is described and illustrated in chapter 1.

Much of the understanding of the concepts of energy is available to readers who have taken courses in basic physics or science. But that does not represent the majority of our population. The subject is important because the uses and sources and future of energy have already become subjects of world concern. Our lack of understanding has started to affect our health, societal behavior, and financial comfort. With a professional background in physics, civil engineering, rocket design, and applied mechanics, the author has produced this book to enable lay readers to understand just what "energy" signifies and how we use it on a worldwide basis without any concern that there will ever be a shortage or a problem.

Concern is sometimes expressed that world energy is being "used up." What does the expression "using up energy" mean? Much of today's world, from governments to individuals, is obsessed "energy" and the way that it is being "used up." Two objectives of this book are to show that energy is the basis for all of life's activities and to show that energy will not disappear. Society's efforts must be focused on using energy efficiently, or converting energy sources to forms that suit our immediate purposes.

In this book we evaluate and quantify the world's principal energy usage (food, heat, light, transportation, and industrial processes). A sensible and economically viable plan is then proposed and described for meeting all our energy needs.

As we consider and examine various energy sources that humankind has available, we can identify the most dramatic and important source for a sensible US (or world) energy policy: *direct solar radiation*. We would satisfy all US energy requirements if we were able to capture and make use of one-hundredth of 1 percent of all the solar energy intercepted by the earth. In this book, which concentrates on a sensible and economical energy policy for the United States, we recommend a policy for providing all our required US energy use.

Our society should realize that almost all of our current energy needs are provided for by limited fossil fuels that will soon disappear, and over half of that supply is dependent on foreign supplies and companies whose interests clash with our national benefits and goals. If we choose to adopt the policies proposed in chapter 7, the issue of an "energy crisis" disappears.

The energy goals described herein are achievable and are summarized in tabular and descriptive form in chapter 7. All we need is creative vision, governmental support, and the determination to become energy-independent soon, and our sensible goals will be achieved completely and economically.

In developing and proposing the policies described in this book, the author's goal has been to avoid complex technological terms and to present to the reader an explanation and policy that is understandable and in everyone's best interests. For this reason, this book includes a number of appendices, primarily for those readers who wish to see further discussion of the material that has been presented. These appendices may be skipped over by the reader who is not interested in mathematical or engineering details. Even so, the policy that is presented here will avoid future energy crises and will lead to a healthier, happier, and economically safer world society.

Chapter 1—Energy and Its Various Forms

A. What Is Energy?

One of the dictionary definitions for "energy" is "the capacity for doing work." In classical physics, "work" is defined in elementary terms as "force times distance." In common usage, the word has more general definitions, but in most of these definitions, it still involves the transmission of forces over distances. Hoisting a weight, loading a truck, pushing a wheelbarrow, and digging a ditch are all manifestations of the classical "work." Subtler examples, but work nevertheless, include pushing a piston, rotating a turbine, walking up a flight of steps, and paddling a canoe.

Energy is therefore defined by me as *a quantity that has the capacity to be converted into work*. Energy may reside in a coiled spring, a pot of hot water, a charged wire, a raised weight, or an unburned fuel, but if it can be used or processed in such a way as to do work, we have an energy source that can be treated quantitatively in terms of the amount of energy it possesses. Energy is essentially the basis for all of life's activities. We use energy to eat, to sleep, to move, and to think. Work and energy are measured using the same units. Whenever work is performed, the energy source is diminished by the same number of units.

B. The Many Forms of Energy

We generally think of energy in two different types of categories: the first is energy in transport, as it moves from one physical place to another; the second is stored energy, in which the energy is contained in a form suitable for release and used at will. In the latter case, the energy-containing material is known as an energy source. These sources are described in some depth in chapter 3. Here, however, we consider nine <u>forms</u> of energy that are currently being used. These nine "forms," as distinguished from "sources," are listed here and then described individually:

1. Kinetic energy
2. Thermal energy
3. Chemical energy
4. Electrical energy
5. Radiant energy
6. Sound
7. Stored mechanical energy
8. Gravitational potential energy
9. Nuclear energy

In the descriptions of these forms of energy, references are periodically made to elements, compounds, atoms, and molecules. These are briefly defined in Appendix 1.

1. Kinetic Energy

This is the energy contained within an object or mass moving from one location to another, such as a bowling ball in motion, an automobile in motion, a hammer in motion about to strike a nail, a piston moving within a cylinder, or a spinning wheel. Bullets kill people because of their kinetic energy. Much of the kinetic energy in the head of a moving golf club is transferred to the ball, which enables it to sail so gracefully into that distant sand trap.

If energy is contained within a material (such as in thermal or chemical form), the act of transporting such material represents a transmission of energy. Examples may include a load of wood being carried into the house, oil being shipped through a pipeline, or a thermal updraft moving along the side of a mountain. On the other hand, kinetic energy is transferred from one mass to another through the direct collision of the masses. These masses may be relatively large, such as football players or billiard balls, or they may be very small, such as atoms and electrons. When these masses collide, causing one to slow down its motion while another speeds up, the energy is transferred. Sometimes the energy is transferred from one large mass to many small ones, such as when a weight is dropped into water and lands on the bottom, or a moving block slides to a stop on the floor. In both cases, the kinetic energy of the large mass has been reduced to zero.

However, as a result of direct collisions with the molecules of the water and the floor, their individual molecules have correspondingly increased in their own kinetic energy. These molecules are too small to see, but their energy can be felt in the form of heat, by measuring the temperature of the water or the floor. This energy is now in the form called thermal energy.

2. Thermal Energy

Some of the nine forms of energy listed previously have the characteristics of motion (such as kinetic energy). Some forms of energy can also be stored and then mobilized only when needed. Such stored energy is sometimes known as *potential energy* because the potential for use is there when required. One common way of storing energy is in the form of heat, also known as thermal energy. However, we can think of thermal energy as a form of kinetic energy. At the lowest temperatures imaginable (minus 273 degrees Centigrade or minus 460 degrees Fahrenheit), the molecules of all substances are motionless, as are their atoms and electrons. As the temperature begins to increase, the electrons, atoms, and molecules become agitated and begin to rotate, vibrate, and collide. This frenzied motion continues to increase with temperature, although it is at such a microscopic scale that we cannot readily observe the changes in motion. But our thermometers and our skin sensors do feel the phenomenon, and we register the change as heat. As the temperature of a substance increases, so does the kinetic energy of its constituent particles. Significant examples of thermal energy include heating blankets, warm air, and a pool heated by the sun.

3. Chemical Energy

Another means of storing energy is in chemical form. Chemical energy is related primarily to the tiny forces and electromagnetic fields that exist between the molecules, atoms, and electrons that make up matter. Electrons are bound to the atom by a certain amount of binding energy, which is much greater for inner electrons than for outer ones.

Similarly, we have interatomic forces and intermolecular forces, all of which can be extremely complex. For example, the electric forces between closely spaced molecules can be attractive or repulsive. If the forces of attraction did not exist, the molecules would separate, and all substances would fall apart. If there were no forces of repulsion, the molecules would all draw together and annihilate each other. In equilibrium, a balance is struck at the distance at which these opposing forces balance each other. Theoretically, the two molecules could now be at rest, perhaps a hundred-millionth of a centimeter apart. If they are moved slightly toward each other, the repulsive forces drive them apart. As they separate, the repulsive force decreases, and the forces of attraction take over and tend to bring them together. As a result, the molecules vibrate toward and away from each other, like two weights connected by invisible springs.

The combination of these intermolecular forces and the phenomena that give rise to them, as well as the force fields that exist at the smaller particle levels, such as atoms, electrons, and particles within the nucleus, are collectively known as *chemical energy*. Chemical energy is a prime component of plants, food, oil, and an unlit match.

The most common form of stored chemical energy is created by photosynthesis. This process refers to the absorption of light energy from the sun by the molecules of chlorophyll contained within green plants. This leads to the production of organic substances such as sugars and starch. Some of the terms associated with photosynthesis are defined in Appendix 2. As a result of this energy absorption, all growing plants act as storehouses of energy within their molecular configurations. Combining the energy of light with water and carbon dioxide leads to the production of food. Animals get their energy by eating plants, and people get their energy by eating the plants and/or animals. When the plants become trees, we call the energy source wood. Over a long period of time and environmental conditions, the wood can become coal. Under different conditions of temperature, pressure, and time, tiny plants and animals can develop into oil, gas, and other fossil fuels. We can remove the stored energy from a fuel by burning it, in which case the energy is converted to the form of heat and from there to light, motion, or some other form. In chapter 4, there is some discussion of the common battery and fuel cells, representing forms of chemical energy that can be converted to electrical energy.

4. Electrical Energy

Energy may also be characterized by the flow of electric current, which is represented by the flow of electrons along a conductor. Electrons are tiny particles of matter that carry a quantity of energy in the form of a negative charge and could be said to revolve about the center of every atom (see Figure 3C-1). If an electron is removed from an uncharged atom, the rest of the atom has a positive charge and will seek an electron to fill its atomic structure and restore its overall charge to neutral. The force with which electricity is moved is called electromagnetic force and can be measured in volts. The electric current itself is a form of energy. As the electrons move through a wire or other conductor, heading toward a positive charge, their passage is partially blocked by the particles that make up the matter, such as those in a wire, and this resistance, just as in the case of friction, results in the production of heat energy, increasing the temperature of the material. Similarly, some electrical energy traveling through a light bulb filament is converted to a form of radiant energy called light (discussed in the next section); if the voltage forces the current through an electric motor, the motor begins to rotate, and the electric energy is converted to kinetic energy; if the current is used to charge a storage battery, the current is converted to chemical energy.

In addition to electrical energy running through a wire, this kind of energy may also be represented by a stroke of lightning.

5. Radiant Energy

Energy may be transmitted through radiation or, more specifically, electromagnetic radiation, which pervades the space all around us. These invisible waves of radiation emanate from many sources and travel at the speed of light until they strike something that absorbs them, reflects them, or permits them to pass through. Radiation may sometimes be thought of as a stream of photons, which are tiny high-speed particles that contain energy but no mass. One can also think of radiation as waves, with wavelengths varying from a fraction of an inch to many miles in length. The longer the wavelength, the smaller is the wave's frequency of transmission.

Electromagnetic waves are characterized by their wavelengths or frequencies and fall into six general bands, which together make up the electromagnetic spectrum: radio waves, infrared, visible, ultraviolet, X-rays, and gamma rays. These waves all carry energy, even though they are of different lengths and frequencies and react differently with matter and the environment.

Special instruments, such as cosmic or gamma ray sensors, X-ray detectors, or radio receivers, are required to catch and record the process. In addition, some radiant energy can be perceived by human beings without any artificial instruments because our bodies have built-in receivers and detectors that react to certain types of radiant energy. One of these types is infrared, which is the energy emitted from a source of heat. Another significant amount of energy falls into the visible portion of the spectrum and is known as light. We sense this heat and light because our skin and eyes have special detectors that are sensitive to these forms of energy and register them in our brains.

Ultraviolet (UV) radiation is not perceived by human sensors, but is a form of radiant energy that is absorbed in the skin, and causes suntan or sunburn.

6. Sound

Although sound also travels in waves, it differs from the previously listed forms of radiant energy because the other waves can travel through a vacuum, whereas sound is dependent on the existence of a medium such as air. When a body vibrates, the oscillation causes periodic waves. These waves travel in the form of a progressive disturbance of air molecules, and so sound cannot exist in a vacuum. The effects of these waves can be detected by sensitive receivers, such as ear drums. The energy transfer, such as from a musical instrument to the air to the human ear, is slight but does exist.

7. Stored Mechanical Energy

Much of the discussion thus far has related to energy in motion, but energy sometimes may be stored without having any moving parts at all. One example of stored mechanical energy is a coil spring that is compressed or extended from its equilibrium state, thereby containing potential energy that can be converted to kinetic energy if the spring is released. A second example of stored mechanical energy is found in a stretched rubber band. The most common examples of potential energy are the instances in which the energy storage is based on the existence of gravity, such as a sled at the top of a hill.

8. Gravitational Potential Energy

A body may possess a great capacity for doing work because of its position and condition, such as a rock sitting on top of a cliff or a sled waiting at the top of a hill. These are two examples of gravitational potential energy. The gravity force between any two particles of matter is sophisticated and complex, and the behavioral reactions of bodies in motion are not considered here. However, we must recognize that every item of mass on earth is being subjected to a vertical force (commonly known as weight) and that if the mass is moved away from the earth and then allowed to fall freely, the mass will generate an increasing amount of kinetic energy by accelerating. Before the mass starts moving, this energy is stored within the mass as potential energy.

The eight forms of energy discussed so far are governed by a physics principle known as the law of conservation. This law states that energy may be transferred from one form to another, but the total amount remains constant (even though different units are used to describe energy quantities, as described in Appendix 3). As an example, consider a common battery that transfers chemical energy to electrical energy. Or as a more drastic example, consider that some of the radiant energy from the sun is absorbed by plants and trees as chemical energy and over millions of years may be converted to a different chemical energy in coal. If this is then burned in a power generating plant, the energy becomes heat and then boils a fluid, causing an armature to rotate (kinetic energy), leading to electric energy that flows through wires into houses, where the energy is then converted into light, heat, and motion. During these processes, no energy is created or destroyed, but the energy does change forms a lot in the way it is used.

9. Nuclear Energy

Unlike the previous eight forms, nuclear energy is created anew and is based on Einstein's 1905 special theory of relativity, which shows that energy can be converted to mass and that under very special conditions a tiny bit of matter can be turned into a tremendous amount of energy. Technology has now developed to the stage in which we can consider nuclear energy as one of our potential sources of energy.

Along with the recognition in 1905 that mass could be converted to energy and vice versa, a new law of conservation was developed. This law states that the total quantity of mass plus energy cannot be created or destroyed, but that one can be converted to the other.

The nine forms of energy previously listed should not be confused with *sources* of energy (e.g., fossil fuels, sun, etc.). For example, electricity is a very convenient *form* of energy that can be easily transferred, converted, and used in many applications. However, electricity is not a primary source for humankind and probably will not be in the future unless we figure out a way to capture and utilize lightning. Since we have not yet done this, we currently obtain our electricity in the United States by converting energy from the following basic energy sources: coal, 52 percent; nuclear plants, 20 percent; natural gas, 15 percent; hydropower, 7 percent; oil and others, 6 percent.

C. The How and Why of Energy Transfer

1. The Transfer of Heat (Thermal) Energy

Energy that exists in the form of heat travels from its source to neighboring (or far) places that are not as hot. When two bodies are at the same temperature, neither one will transmit energy to the other. However, if they are at different temperatures, an energy transfer will occur, tending to equilibrate them. This thermal energy transfer can occur in three ways: first, if the bodies are touching each other, heat is transferred directly from molecule to molecule, and the process is known as conduction; second, if the bodies are near each other but not touching, currents of air can carry the heat from one body to another through convection, which is equivalent to conduction except that the air (or water, etc.) is used as an intermediate medium; and third, if the bodies are separated with no intermediate medium, radiation is the means by which energy is transmitted away from the hot body.

Soldering irons and electric blankets transfer heat through conduction; the Gulf Stream, baseboard heaters, and double boilers are primarily dependent on convection; the sun, the stars, and heating lamps transfer their energy through radiation. In every case, though, the direction of energy movement is from the hotter body to the cooler one.

2. Energy Transfer in Chemical Reactions

Whenever a chemical reaction takes place, the chemical process may absorb or release energy to and from the atmosphere, in which case the environment can be cooled or heated by the process. Throughout nature we can find various combinations of materials that may themselves be in equilibrium but that, when permitted to combine or separate under the right conditions, undergo chemical restructuring with an associated release (or sometimes increase) of energy. For example, when hydrogen (H_2) and oxygen (O_2) are permitted to combine at room temperature, water (H_2O) is formed. Since the chemical energy of water is much less than the chemical energy of its constituent hydrogen and oxygen when separated, the formation of water is exothermic—that is, energy is released. Similarly, if pure nitrogen (N_2) and pure hydrogen are mixed, they combine to form ammonia (NH_3), but much less energy release accompanies this reaction.

Combustion is the name given to chemical reactions that occur rapidly and give off heat and light—known as an exothermic reaction. Generally, combustion (burning) occurs when a substance combines rapidly with oxygen. The process of combining with oxygen is known as oxidation. In the case of rapid oxidation, the combined chemical energy content of the oxygen and the combining substance (fuel) before the combustion occurs is much greater than the chemical energy in the products of combustion. The difference is radiated away as heat and light. Examples of slow oxidation are the rusting of iron and the production of carbon dioxide in the body, which does not give off the heat and light needed for combustion.

Commonly, the fuel in combustion is a hydrocarbon or carbohydrate, such as paper, wood, rags, oil, coal, or methane gas. After efficient burning, the primary products of combustion are carbon dioxide and water. The energy levels of these two substances are quite low, and there is little likelihood of their entering into a further exothermic reaction.

3. Photosynthesis

In later chapters of this book, we discuss various effects of solar energy. A primary phenomenon of life, triggered by solar radiation, is known as photosynthesis. Within green plants there are molecules of carbon dioxide (CO_2) and hydrogen-containing pigment molecules known as chlorophyll. Within each of these molecules resides the usual turmoil of atoms, electrons, neutrons, and superparticles, all of which are tugging, repelling, vibrating, and spinning, with resultant well-defined energy levels. In the meantime, the sun, which is a hot body, radiates heat away in all directions and at all wavelengths. Some of this energy, in the form of photons of visible light, travels to earth, gets through the atmosphere, and strikes the chlorophyll molecules in the plants. When this happens, a chemical reaction is triggered, and the elementary particles within the plant are rearranged. Some carbon dioxide is created and absorbed, some oxygen is released, some plant growth results, and the total chemical energy within the plants is increased. This leads to the fundamental supply of food that sustains life on earth.

4. Radioactivity

One further phenomenon that produces energy change is radioactivity. A radioactive substance is one that is not in equilibrium. Just as in the case of mechanical systems, nature seeks to establish atomic structures that are stable. Yet there are unstable atoms that have been created through various complex processes and that tend to transform toward an equilibrium configuration. During this transformation, which might take microseconds or eons, there is a periodic restructuring of the atom with an attendant reduction in stored energy. This excess energy is released as radiation.

We see, therefore, that energy transfer can occur in many different ways. In the quest to improve the quality of life on earth, we take advantage of these phenomena to create desirable energy transitions. On the other hand, sometimes these phenomena are too big for us to manipulate, and humans must then learn to live with the energy environment as supplied by nature.

D. Some Examples of Energy Conversion

1. Einstein's Exception

The world around us is made up of matter and energy. At the beginning of the twentieth century, two fundamental principles were taught to every student of elementary physics, namely, the principles of conservation of matter and conservation of energy. For example, the principle of the conservation of matter stated that matter cannot be created or destroyed; it only can be changed from one form to another. Ice can become water, and water can become steam, but the weight of the constituents and the total number of molecules remain constant. When a piece of wood is burned, it is converted into ashes and various gases, but the total mass is unchanged.

But we were wrong. Between 1905 and 1923, Albert Einstein revolutionized the scientific world by showing that matter and energy are interchangeable rather than distinct. His famous equation, $E = Mc^2$ (energy equals mass times the square of the speed of light), showed that matter could be destroyed, as long as it was converted to its equivalence in energy. However, it takes a very tiny bit of matter to produce an extraordinarily large amount of energy. This was the basis behind the concept of the atomic bomb and the peaceful uses of nuclear energy.

2. From One Form to Another

A good example of energy conversion lies in the use of a piece of coal to toast some bread. Having decided that we want our bread toasted, we must now apply some heat to it. We have available a piece of coal that can be burned to convert its chemical energy into heat, most of which will be lost to the atmosphere, but some of which can be used to toast four or five slices of bread. Instead of using it for this purpose, we ship the coal to a power generation station. There it is burned, giving up most of its energy to the atmosphere, and some of it to the heating of water and production of steam. Some of the steam heats the atmosphere, and some of it expands against a turbine blade, contributing toward its rotation. The turbine drives a dynamo that helps heat the atmosphere somewhat and simultaneously produces some electricity. The electricity is transformed, stored, and sent along transmission wires that all further heat the atmosphere while leading to a power source at your kitchen outlet. You now plug a toaster into that outlet, and the electrical energy is converted into heat, some of which is used to toast one slice of your bread. The rest of the energy goes into the atmosphere as heat.

Instead of getting enough energy out of the coal to toast five slices of bread, we diverted it sufficiently and circuitously enough that we ended up with only one slice of toast. The rest of the energy has gone into the atmosphere, from where it radiates into space. But we *did* get our toast.

Another example of this principle is demonstrated in tracing some of the energy from the sun when it intercepts the earth. As described previously, a photon of visible energy from the sun is absorbed by the chlorophyll molecule within a growing plant that then stores the energy in chemical forms such as sugars and other carbohydrates, simultaneously releasing some oxygen. When the plant dies, its fiber and leaves slowly disintegrate, absorbing oxygen and reversing the chemical photosynthetic process, which releases the stored chemical energy into the atmosphere as heat. If the plant is buried so that no oxygen is available to combine with it, the chemical energy may remain stored within its molecules for a long time, while pressure and heat may change its form. Thus, we see a spate of "fossil fuels"—oil, coal, coke, shale, gas—all of which contain the solar energy that was captured when the photon struck the chlorophyll molecule. In every case, the ultimate disposition of this energy is to heat the atmosphere after it is released from the fuel (upon combining with oxygen) and then to radiate away from the earth into space. We see, then, that the energy has not been created or destroyed. It started as radiation from the sun and was successively transformed into stored energy in a living cell, stored energy in a fossil fuel, heat in the atmosphere, and radiation into space.

A second transition path for stored solar energy within a living plant occurs if the plant is eaten, after which it may be metabolized within the body as the energy permits the eater to do work (breathe, walk, jog, lift weights, etc.); it may remain as chemical energy within the body until eventual death and decay; or it may serve to keep the body warm until it radiates away. No matter what path it takes, however, the solar energy of photosynthesis is traceable and its residence on earth is temporary.

As we begin to examine energy sources and energy uses in the pages to follow, bear in mind that an energy source merely stores energy. The energy came from someplace, and it will go someplace. Eventually, it will heat the atmosphere and radiate away to space. In the meantime, new energy will be received from the sun.

E. The Earth's "Energy Balance"

Imagine an empty can with a small hole near the bottom. Using a hose, we begin to pour water into the top of the can, and the water escapes through the hole. If the flow of water from the hose is greater than the rate at which it can escape from the hole, the level will rise, and the total amount of water in the can will increase. If the hose is shut off, or the flow is reduced so that the water flows out faster than it enters, the level will drop, and the can will tend to empty. Finally, if we adjust the hose so that the amount of water entering is just equal to that which is escaping, the water level remains constant, and we have achieved a condition of equilibrium.

The can and the water are analogous to any substance or body that is subjected to the input and output of energy. The body may be a bacteriological cell, a flower, an electric appliance, a house, an ocean, or a star. When the amount of energy that it absorbs in any period of time is equal to the amount that it emits, the body is said to be in energy equilibrium, and its average temperature remains constant. If more energy enters than leaves, its temperature increases, and it gets hotter. If the energy that exits exceeds that which enters, it cools off, and its temperature decreases.

The difference between the energy that comes in and the energy that goes out represents the additional amount that is stored or removed from storage. The calculation that compares all of these so that all energy is accounted for is called an "energy balance."

Generally, all structures and materials tend toward a condition of energy equilibrium, but sometimes it takes a long time to reach this condition. For example, a radioactive material undergoes internal physical and chemical changes that convert electrical and atomic energy to heat, which then radiates away from the material. This process, known as "radioactive decay," may take a long time. If we observe the material, make suitable measurements, and conduct an energy balance, we find that the energy (heat) that is emitted is greater than that which enters. We conclude that the material must be releasing its own stored energy, and we classify the substance as an energy "source."

Sometimes our energy balance shows that the heat entering a body is greater than that which leaves and that the excess energy is obviously being stored within the body in some form, in which case the substance is known as an energy "sink."

Let us now examine the earth as a single body on which we will conduct an energy balance. By "earth," we refer to the spherical body with which we're all familiar, consisting of a solid crust, a mantle, a hot liquid core, oceans, and an outer layer of gas varying from a dense atmosphere to a rarified ionosphere. It is more than 8,000 miles in diameter, and it spins around its own axis once a day, revolves around the sun once a year, and hurtles through space as part of the solar system.

A measure of the average sea-level temperature at the surface of the earth produces an interesting result—that is, it shows that it is not changing. Scientific observations show that this average temperature has been approximately 59 degrees Fahrenheit for many years. ("Average" takes into account the variations throughout the year and over the surface, including the equator and the poles, the deserts, and the forests.)

Over the earth's four-billion-year history, its structure and temperature have obviously changed significantly, but these changes have leveled off greatly, and although there have been many cycles of "global warming" and "global cooling" in the last few million years, an energy balance for the whole earth shows that it is essentially in equilibrium; the energy going out is equal to the energy coming in, and the earth is neither heating up nor cooling off dramatically. Primarily as a result of cycles in solar activity (such as a recently observed 1,500-year cycle), there have been, and will continue to be, regular periods of "ice ages" and "global warming," but the average surface temperature of the earth will not vary from its current average of 59 degrees Fahrenheit by more than a few degrees.

Looking first at energy input, we must realize that there is essentially only one outside energy source that provides the earth with its heat and light, and that is the sun. Starlight and other space energy sources are negligible when compared with the sun. This solar energy falls on one half of the earth at any single time, but as the earth rotates and night changes to day, and as the earth's spinning axis wobbles about the sun, each portion of the earth receives some solar energy in its turn. The energy falls most directly on the equatorial regions, whereas it strikes the polar regions only obliquely.

Since the sun is very far away from the earth (93 million miles) and radiates in all directions, a very small portion of its radiation energy is intercepted by the earth. Only two-billionths of the total energy radiated from the sun reaches the earth. Let us now consider what happens to this energy when it gets there.

Section C-2 of this chapter refers to the three ways in which heat energy is transferred: conduction, convection, and radiation. Of these, it is obvious that the only mechanism for energy escape from the earth is through radiation. Convection may cool the surface and carry some of the energy into the atmosphere, but since there is no air to carry it further, it can escape into space only by radiating away. The amount of radiation energy depends on the temperature of the radiating body. Near the equator (between North and South 38 degrees latitude), more energy is received than is emitted, so that there is a net heat gain, and the temperature rises above the world average. Similarly, the polar regions undergo a natural heat loss since they do not face the sun directly and radiate away more than they receive. The temperature extremes that this might be expected to produce are minimized, however, by the averaging effect of the blanket of air around the earth. Natural convection combines with earth spin and gravity to produce ocean and air currents that distribute weather over the globe, and so the temperatures at the poles are greater than they would be if we had no atmosphere, whereas the temperatures at the equator are less.

Combining all of these effects, the earth radiates energy into space as though the temperature were uniformly 59 degrees Fahrenheit over the entire surface. If we now balance this energy with that which is received from the sun, the net effect is to produce an equilibrium condition over any meaningful time period.

The energy balance for the earth is pictured in Figure 1E-1, where we see a graphic representation of what happens to the energy from the sun when it reaches the earth and its environs. We can see that much of the energy that reaches the vicinity of the earth never gets absorbed into the solid surface. Nineteen percent is captured in the atmospheric gases above the earth; some of this energy forms the ionosphere, some is absorbed by oxygen and ozone, and much is absorbed by water vapor and clouds.

An additional 34 percent of the solar radiation doesn't get absorbed anywhere, but rather bounces off the cloud and earth surfaces as if they were mirrors and is reflected back into space. This is not a surprising phenomenon for energy in the solar spectrum, which is primarily in the form of infrared, visible, and ultraviolet electromagnetic waves, as discussed earlier in this chapter.

The remaining 47 percent of solar energy is absorbed at the earth's surface, where most of it heats the land and the oceans and tends to flow from the surface inward. A portion of this energy directly heats our bodies, rooftops, and solariums. A significant portion of it is captured by forest and plants, where it undergoes photosynthesis and is converted to a stored form.

Because of the earth's surface temperature, the energy that is radiated away is initially greater (115%) than the solar energy that reaches the earth. In addition, air that is warmed by the earth and carried upward removes an additional 10 percent from the surface. Finally, through the cycle of ocean water evaporating, rising as a vapor, and then condensing and falling to earth as rain, an additional 19 percent is carried from the earth's surface into the atmosphere. This outward-moving total of 144 percent is offset by 97 percent that is reflected back from the atmosphere, resulting in a net outflow of 47 percent. Thus, the net energy emitted is equal to the net energy absorbed.

This presentation of many different outgoing and incoming energy quantities can be better understood by referring to Figure 1E-1 while reading the foregoing paragraphs. It will be noted in this figure that the total energy from the sun that intercepts the earth is shown to be $5.4(10)^6$ quads/year (or 5,400,000 quads per year).

A "quad" is a very large amount of energy that is referred to and discussed throughout this book. Just as weight can be expressed in pounds, ounces, grams, tons, kilograms, and so on, quantities of energy can be expressed in many ways, some of which are listed in chapter 2, section A, and Appendix 3. The term "quad" is a convenient way of expressing a quantity of energy, and we use it to discuss world energy sources and availability. For example, even though the world uses about 450 quads per year, this is a small number compared to the annual received energy shown in Figure 1E-1.

One should consider the effect of the earth's interior, which is known to be a hot molten core. Does the earth behave like a baked apple, which is slowly cooling off at the center as the heat is transported to the surface? The best present estimates are that it is not. Instead, it is believed that there is a significant amount of radioactive material in the region of the earth's mantle, generating a quantity of heat that keeps the core at a constant temperature in the vicinity of 5,000 to 9,000 degrees Fahrenheit. Although some heat seeps up to the surface and is absorbed by the oceans and atmosphere, it is far less than one-tenth of 1 percent of the incoming solar radiation, and it is insignificant in the overall energy balance of the earth.

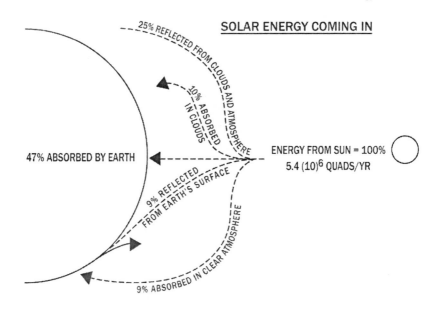

SOLAR ENERGY COMING IN

25% REFLECTED FROM CLOUDS AND ATMOSPHERE

10% ABSORBED IN CLOUDS

47% ABSORBED BY EARTH

ENERGY FROM SUN = 100%
5.4 (10)^6 QUADS/YR

9% REFLECTED FROM EARTH'S SURFACE

9% ABSORBED IN CLEAR ATMOSPHERE

ENERGY LEAVING EARTH

19% LEAVES BY MOISTURE EVAPORATION

115% LEAVES BY RADIATION

TO SPACE

(47% NET EMITTED)

10% LEAVES BY CONVECTION

97% REFLECTED BACK FROM ATMOSPHERE

Figure 1E-1

The Earth's Energy Balance

F. What Is Meant by "Using Up" Energy?

In the international discussions of the world's energy problems, reference is commonly made to the fact that the industrial nations use energy at a much greater rate than third-world countries and that the United States is profligate in its use of energy. But we have just seen that energy is not destroyed or "used up," but rather is converted from one form to another. When we speak of the use of energy, we really refer to the way that society takes advantage of a quantity of energy in the form in which it happens to be found, or changes it into some other form in which it meets a societal need. Following are examples of this:

- An individual chooses to travel to some other location a significant distance away. If he walks, it will take a long time, and he would like to get there fast. Therefore, he converts energy in a stored form (gasoline) to generate heat that drives an engine, which goes into motion of the vehicle (kinetic energy), which ultimately heats the atmosphere and radiates out to space. During the period that a portion of the energy is in the form of vehicle motion, he takes advantage of it and is carried to his destination.

- An individual would like to swim in the ocean, but she is comfortable only if the water is in a compatible temperature range. Solar energy radiating from the sun falls on the ocean surface and is absorbed as heat and then radiates away into the night sky. During the period that the water temperature is comfortable, she gets her swimming done.

- An industrial company that produces specific metallic shapes (frying pans, auto fenders, or ingots) needs temperatures high enough to melt the raw materials so that they can be cast or formed. The energy can be purchased in the form of electricity, or in a stored fossil form such as coal or oil or gas. This energy is converted into heat that enables the metal to be formed, after which it is cooled and the heat travels into the air or a coolant material.

In each of these cases and in every case of energy use, including body sustenance and human comfort, people take advantage of available energy in a convenient form for their purposes, but they do not create the energy, and when they have completed their enjoyment of it, the amount that they have "used" goes elsewhere without diminution.

If energy is not "used up," one might ask, does it then go on forever in the universe, changing from one form to another, but always available for humans (or any other creature) to convert and exploit for their purposes? The answer is that energy is not always available. Certainly such a concept would violate our fundamental and intuitive understanding that there are no perpetual motion machines or free lunches in nature; for each action we take, there is a price. In this case the price is an increase in the unavailability of energy:

We have seen that energy travels under certain conditions. For example, when a hot fluid is mixed with a cold fluid, the heat flows from one to the other until a uniform temperature is reached throughout the mixture. This energy transfer occurs without any external stimulus. But once the temperature has reached equilibrium, no further heat flow occurs, and even though the total amount of energy has not changed, we can no longer expect the condition to restore itself to half hot, half cold. The energy in the hot fluid was available to flow into the cold fluid; it is still in the mixture, but it is not available to flow back.

This proclivity for uniformity is found in every natural phenomenon. When two pressure zones meet, a wind is created and tends to equilibrate them; the surface of water seeks a uniform level; and the interfaces of solids and fluids tend to become less distinct with time. Each time such a phenomenon occurs, there is a lessening of extremes, but no change in total energy.

This leads scientists to conclude that the ultimate disposition of the universe is to develop toward a condition in which the energy will no longer be available for transfer via motion, conduction, or radiation. There will no longer be discrete bodies in space. The universe will consist of a colorless, constant-temperature, uniform continuum. Fortunately, this is some time away.

And so when we speak of "using up" energy, we really mean using up fuel, or converting energy sources to some form that suits our temporary purposes.

Chapter 2—The World's Energy Use

A. Energy Measurement Units

In chapter 1, the discussions of energy were limited to definitions, descriptions, concepts, and categorizations. In this chapter, we define the various units of energy and the mathematical equivalence of these units.

The quantity of energy used for any activity varies from a tiny bit (e.g., the amount contained in a moving electron) to a massive quantity (e.g., a tsunami, or the birth of a new star). Therefore, a large number of terms are used to describe energy quantities, depending on the applications. Although the rest of the world uses the metric system of measurement, the United States continues to use the English system, although it is more complicated and less logical. However, when considering energy measurements around the world, where we are dealing with distance, mass, weight, force, volume, liquids, solids, amounts of electricity, sizes of land areas, and bushels of agricultural material, different societies and industries use their favorite and conventional terms, and the reader of an energy document must be prepared to convert quoted units to the form with which he or she is familiar. For this reason, Appendix 3 features a listing of the units used by physicists when they discuss energy, as well as the conversion factors in switching from one form to another. Many of these common units are not used in the discussions of energy in the following chapters. However, some of them are interesting and valuable for the reader to know. Those that are important for the discussions in this text are listed here, as well as being repeated in Appendix 3:

BTU or *Btu* (*British thermal unit*): This is a principal energy unit used by mechanical and thermodynamic engineers. A BTU is the amount of energy that will raise the temperature of one pound of water by 1 degree Fahrenheit. This energy could be introduced into the water mechanically, such as by stirring, or directly through a heat source.

Quad: This is a convenient term that has become popular in describing energy needs of entire countries, or large portions of the world. A quad is one quadrillion (1,000,000,000,000,000) BTUs.

In this book we describe the energy content of various fuels using some of the preceding terms, but concentrating principally on quads. However, it is convenient to present other conversion tables that relate the equivalent energy values between coal, gas, and oil, as well as their absolute values in BTUs. For example, many documents that discuss energy sources use different units in quantifying energy; they may describe oil in terms of barrels, coal in terms of tons, gas in terms of cubic feet, electricity in terms of kilowatt-hours, and so on. In discussing the energy contents of these different sources, one must resort to very large numbers, such as billions, trillions, quadrillions, and quintillions. To avoid the confusion that this can cause, in this text we regularly convert these quantities to quads and discuss current availability and future needs in terms of this single parameter. In addition, the tables in Appendix 3 enable the conversions to be made more easily, as well as provide further energy definitions. These tables can be used to determine how one type of fuel can be substituted for another to provide the same energy capacity. For the reader's convenience, quads of energy in terms of common fuel sources are shown again here:

1 quad = 38.5 million tons of coal = 172 million barrels of oil = 971 billion cubic feet of natural gas = 60 million dry tons of biomass (agricultural residue).

B. World Energy Consumption (by Use)

As described in chapter 1, every universal and earthly phenomenon involves the use of energy. This is discussed and treated in many ways in textbooks, classrooms, and reference sources; however, no comprehensive analysis has ever been made that quantifies the total energy dissipation or usage on a global basis. This is logical and justifiable, since there is no reason to require such a universal justification, and it would be very hard to do. For example, who cares how much energy is expended by an ant colony building a nest? Or by a large tree branch breaking off during a windstorm? Some energy exertions are natural occurrences that no one thinks about. Others require human planning and action to divert and/or apply the required energy. In these latter instances, people must categorize the energy form, establish measurement systems, and introduce them into society's economic structure.

Chapter 1 lists nine different forms of energy, and chapter 3 describes the different primary sources that can be used to provide these various forms. However, in considering how humanity uses the available energy, we can break this usage down into four categories: to provide food, heat, transportation, and industrial processes. In the pages that follow, we see that it is sometimes difficult and sometimes impossible to measure the specific forms of energy that are used in each of these categories, but a general limited quantification is possible and is presented in this section.

1. Food

Various life forms require a number of substances in order to survive and grow. Humans need substances such as carbon and nitrogen to build and repair tissue, keep the body in good working condition, and supply fuel for energy. These substances fall into three major groups: proteins, carbohydrates, and fats. In addition, the body needs vitamins and minerals. In the process of life development and during normal activity, these substances develop into more complex compounds and then decompose as energy is exerted. Ultimately, the life forms die, and the complex cells decompose and decay until they return to their original basic elements. There will have been no creation or loss of matter, but energy will have been used in the process. The substance that provides this energy for living beings is called *food*.

We see, then, that all life forms require energy to produce the necessary chemical changes for cell growth. In addition, higher life forms call for additional quantities of energy, to permit voluntary and involuntary muscular activity and to maintain the temperature at which the life forms can survive. The energy in food is measured in terms of Calories, defined in Appendix 3. (Note that "Calories" representing food energy is spelled with a capital C, as distinguished from "calories," also defined in Appendix 3.) The nutritional requirements of food to sustain normal life must therefore define the chemical substances that are required (fats, carbohydrates, proteins, minerals, and vitamins), the environment (temperature, available air and water), and the Caloric energy levels.

Caloric requirements for human beings depend on sex, age, body weight, and degree of physical activity. The average human being requires about 1,500 to 3,000 Calories per day to thrive. Children require much less; large male laborers may require more. The average US diet provides close to 3,000 Calories per day. Underdeveloped countries are rife with starvation and malnutrition, and half of the world's population ingests fewer than 2,000 Calories daily. Let us assume here that it is an objective to provide an average of 2,000 Calories for all the people on earth. With a current world population of more than 6 1/2 billion people, and using the conversion factors of Appendix 3 (from Calories to quads), we can determine that the food energy that people need to live healthy lives is close to twenty quads per year.

It is interesting to note that this energy requirement for food is almost never counted in the energy inventories that are frequently published. In most analyses of energy supply versus demand, the food energy requirement (in calories) is generally ignored. One logical reason for this is that the food energy use is very small compared to that which is used for transportation and industrial processes, as will be shown shortly.

2. Heat

One of the manifested forms of energy discussed in chapter 1, section B, is heat (thermal energy). More specifically, consider the warming effect required to keep the human body temperature in the vicinity of 98.6 degrees Fahrenheit. The part of the world society that lives near the equator is not particularly concerned with the type or amount of energy required to keep warm; solar energy does it automatically, and no one really worries about it. The thatched hut in the tropics is warmed by the surrounding air via solar energy, but this is not counted in standard energy calculations.

On the other hand, as we consider regions moving north or south from the equator, the body starts to get uncomfortable, and so we wear clothes to provide necessary warmth, which is now achieved by insulating the body from surrounding cold air. In this case, there can be a measurable energy expenditure in manufacturing the necessary clothes, but this can be characterized separately as part of the required energy for industrial processes. When we reach the really cold regions, we may require electricity or gas heating systems to keep us comfortable, and the energy for these can now be measured and classified in various ways.

The Eskimo in his igloo may burn some kerosene to provide warmth; this is almost always counted. On the other hand, if an Eskimo family is keeping its igloo warm by burning the blubber from a whale whom it killed, this energy source is not generally measured or discussed as a normal heating requirement. The native who lives on the fringes of a forest area may burn some wood to give him warmth; this is sometimes counted. However, the solar energy that heats a house in Ecuador is not a measure of concern, and it is taken for granted or ignored. Much of the energy that we receive in the form of "heat" is a natural resource that we obtain at no cost. However, the energy that is clearly paid for to provide warmth (such as gas for heating a home) is in the order of twice the energy needed for food, or forty quads.

Some authorities skip the entire issue by avoiding a classification called "warmth" and instead divide the energy-using agencies into residential, commercial, and industrial, but again there are many overlaps here. "Residential" use of energy generally involves central heat and air-conditioning, hot water heat, food heating and refrigeration, and electric lighting and convenience appliances. "Commercial" and "industrial" users are involved in these activities also, but the emphasis and the distribution are quite different.

And so as we list and analyze world energy resources, we numerically evaluate those that can be measured, such as travel in a gasoline-powered automobile, and we ignore those that are hard to measure, such as personal physical exercises.

These uncounted energy sources represent many hundreds of quads per year. Since we get most of it "free" and don't count it anyway, it is not included in the listings of energy expenditures discussed in this chapter. Instead, we count the energy production that is measurable, such as the outputs of mines, oil fields, utilities, and large production plants; these are included in the discussions of industrial processes.

3. Transportation

It takes energy to travel, and as countries become more advanced, a greater percentage of the population wishes to travel. It starts with local visits, expands into intercity transportation, and finally develops into global coverage. Societies start out by walking, then switch to beasts of burden and self-powered vehicles such as bicycles and canoes, and eventually look for comfortable high-speed vehicles with external power sources. Approximately one-fifth of all the energy that is expended in the world is devoted to moving people or cargo. We obtain the energy in a stored form called fuel, convert the fuel energy to a kinetic form as we accelerate our vehicles, and then change it to the form of heat as we step on the brakes or pull back on the stick.

To transport people or things, we use automobiles, buses, trucks, trains, airplanes, helicopters, barges, and boats. The most convenient way of powering most of these vehicles is to start with chemical energy in the form of a refined petroleum. And so the world has become heavily dependent on oil. In the United States, two-thirds of all our transportation energy is provided by the consumption of petroleum.

As we look into the future (or read further in this book), we can anticipate that transportation energy sources in the future may involve alcohol (such as ethanol), hydrogen, stored electricity, liquids made from coal, biodiesel fuel, and liquefied gas. Without oil we would have to redevelop all our means of transportation to use these alternate energy sources, and this could introduce a significant discontinuity into our social, industrial, and economic structure. Such a discontinuity could be minimized by taking corrective actions now.

In addition to the energy used in the form of fuel for vehicle operation on land, air, and sea, consideration must be given to the energy required for vehicle manufacture and maintenance, operation of terminals and networks, and land transport expenditures (such as roads). Of the transportation means discussed here, the principal user of energy is the automobile. It is assumed here that there are currently about 500 million automobiles and trucks in use throughout the world, with more than half of them in the United States. On a worldwide basis, the typical vehicle is driven about 12,000 miles in a year, and in the process, more than 700 gallons of gasoline are used. This represents a world energy use of 350 billion (500,000,000 × 700) gallons of oil per year. This corresponds to a little more than 8 billion barrels of oil, which is equivalent to forty-six quads of energy per year. Allowing also for the usage of trucks and buses, a reasonable estimate for total transportation energy is fifty quads. It is of interest to note that about three times as much energy goes into automobiles in the form of fuel as goes into people in the form of food.

In the near future, the number of vehicles in the world is expected to continue to increase with population as well as with new economic development in countries such as China and India. On the other hand, the developing automobile technology is expected to produce significant increases in mileage performance, leading to the conclusion that automobile annual energy needs of sixty quads is not an unrealistic estimate. In the near future, we can expect to see continuous improvements in the average fuel economy of conventional cars and trucks. More significantly, there will be some dramatic changes in transportation technology.

The energy to fuel airplanes, trains, ships, and other modes of transportation is taken here to be equivalent to half the amount used for automobiles. This results in a total energy use for transportation of approximately ninety quads.

4. Industrial Processes

Industrial processes are very interesting in that they vary with our societal tastes and development. If it becomes fashionable for each of us to have an electric toothbrush, a heated sauna, or an atom smasher in our home, we will all strive to achieve this and will call for the corresponding energy requirements. For example, we saw previously that the world food energy requirement is approximately twenty quads per year. The energy used to fertilize, irrigate, and harvest such food is of the same order of magnitude. But when we consider satisfying our desires to package, process, cook, can, transport, freeze, modify, and distribute this food, in addition to special boxes, bottles, TV dinners, and attractive displays, the total energy requirement can be quadrupled.

Similarly, we have developed exotic tastes in the way we want to live, all of which require energy expenditures. In addition to eating, keeping comfortable, and traveling, there are other things we want to do. We find it convenient to use electricity. We want to light our cities at night. We want to cast aluminum into pots and forge steel into automobiles.

The fuels that have been used for these various industrial processes are easily measured, consisting primarily of oil, coal, and gas. The annual world expenditure for these many industrial processes is estimated to be in the neighborhood of 300 quads.

5. A Summary of World Measurable Energy (by Use)

The foregoing sections describe a number of world energy sources that are not measured, primarily because they are free, small, or natural without requiring scientific analysis (such as breathing or running around the block). On the other hand, a number of these energy needs are measurable and must be considered as we look into the future and determine what energy sources will be available to fulfill these needs. Combining the values shown previously for food, heat, transportation, and industrial processes, we determine that the total global energy use at the present time is in the order of 450 quads (20 + 40 + 90 + 300). Of this quantity, it is estimated that 20 to 25 percent is used by the United States. Although they use much less than the United States, other principal energy users (in order) are China, Russia, Japan, Germany, India, Canada, and the United Kingdom.

The qualitative values shown in this book are based on current analyses and result in projections that are valid for today's world, so that the values of 450 quads for the world and 100 quads for the United States are realistic estimates of energy use in the first decade of the 21st century. It must be realized, however, that global energy use has been growing rapidly and is very much affected by the increase in population, as well as scientific knowledge and technological advancement. And so we realize that the 30 percent increase in US energy use in the last three decades is primarily due to a significant growth in US population (primarily due to immigration), whereas the energy in the rest of the world has been growing at a slower rate, due to newer and faster means of transportation, combined with increases in life expectancy and scientific knowledge.

Projections for future world energy growth can be somewhat frightening. Official statistics for predicted world energy use are in the growth range of 1 percent per year, but this applies primarily to Organization for Economic Cooperation and Development (OECD) nations, where population and energy growth for buildings and transportation are controlled somewhat by established energy markets. On the other hand, the non-OECD nations (e.g., China, India, Russia, etc.) are seeking and projecting robust growth to satisfy their demands, and we can anticipate that their annual energy use will increase at a greater rate than 2 percent per year.

C. Measurable Energy Consumption (by Source)

At the present time, the distribution of energy sources is heavily influenced by current prices, required level of technology, the shortsightedness of our leaders, and the reluctance of the general population to give much weight to long-range planning. This book proposes a program for energy use that is far different from the one we have today. However, as we make comparisons of what the energy program could be like in the future, it is important to be aware of what it is like today. The following is an approximate distribution of the sources of world energy consumption at the beginning of the 21st century. The percentages in the table are applicable to the world in general and the United States in particular.

Energy source consumption	Approximate % of total
Oil	39
Coal	25
Natural gas	23
Nuclear	6
Wind and biomass	5
Others (hydropower, geothermal, solar)	2

It is of interest to note here that all of these energy sources were originally products of the sun, and the sun is still the basic source enabling almost all of them to function. In the recommended energy policy described in this book, solar energy is shown to be the logical and practical basis of humankind's existence in the next century.

The United States has only about 5 percent of the world population, but it currently uses 25 percent of the present developed energy sources. Another country in a similar position is Japan, which has about 2 percent of the world's population but uses about 6 percent of its currently available energy. It is important for the United States to have a sensible economic policy and an efficient strategy for meeting all of its energy needs. In making comprehensive analyses of future energy needs and supplies, it should be noted that world energy use has been growing at a rate of about 2 percent annually. This rate may be a reasonable one to assume for future growth, although this book is concentrated primarily on the development of a sensible program for *current* needs.

The analysis here is therefore concentrated on the US total annual energy consumption, which is taken here as 100 quads. This energy comes primarily from fossil sources, which will not last very long at their present level. As we examine our energy budgets and the sources that are available, we must look for ways in which the 100 quads can be provided with the least disruption and the most benefit to society.

Chapter 3—Energy Sources: Types and Available Quantities

All of the energy that is available to people on earth comes from three basic sources: first, the largest member of our solar system, the sun; second, one of the smallest particles in the world, the atomic nucleus; and third, stored energy in the form of heat and motion, some of which may be available to us. We shall consider each of these sources in turn.

Up to the nineteenth century, the prime world energy sources were wood, water, and wind. Then came fossil fuels (coal, oil, and gas). As the US economy expanded in the mid-twentieth century, there was a significant increase in the import and use of foreign oil. Soon, most western Europe countries, Japan, and most developing countries became dependent on OPEC-produced oil. It should be recognized, however, that oil is only one of a large number of energy sources. The world uses it because it is presently convenient. But that convenience can easily disappear. In this chapter, we consider the other types of energy sources that could also be available.

So far, this book has been devoted to explaining how energy is defined, what its many forms are, and how it can be converted from one form to another. In this chapter, we identify and discuss seventeen sources of initial energy as presented in the following table. (Interim energy forms such as electricity and pure hydrogen are discussed in chapter 4.)

Fossil fuels	Direct solar radiation
Coal	Solar thermal

Oil	Solar voltaic
Natural gas	Solar energy in general
Oil shale	**Indirect solar energy**
Tar sands	Hydropower
Biomass	Wind
Wood	**Other energy sources**
Grasses	Nuclear fission
Biofuels	Nuclear fusion
Organic wastes	Geothermal

Before making comparisons of these sources, we must recognize that they are not readily interchangeable. Certainly, as far as the potential for storing or releasing BTUs is concerned, they might be compared. But we must also examine how easy it is to effect those releases, what their physical forms may be, and what by-products they will produce. If we want to run an internal combustion engine, an energy source in liquid form is convenient. If we want to power a boiler, easy conversion to high temperature is desirable. If we're going to grow a flower, let's use high-velocity photons.

We must also consider what importance we shall ascribe to the "efficiency" of the energy conversion process. A big deal is generally made of efficiency ratings, but that is because one usually thinks of a specific conversion process derived from a specific energy source for a specific purpose. If the remaining energy cannot be converted for the particular process, it is labeled "waste energy," but it generally does serve some additional purpose. It may heat the atmosphere (and hence our streets), or it may contribute toward the creation of winds and rain.

For example, if we are trying to encourage the growth of grasses and plants, and only 1 percent of the incident solar energy is captured during photosynthesis, the other 99 percent is not necessarily lost. Some of it warms the oceans, causing vapor to rise and become rain; the hydropower to which this leads is not a new energy source but rather part of the 99 percent. Also, because the atmosphere has been heated, we need less of other energy sources to warm our bodies and homes. If we are sufficiently creative and are prepared to use available energy in many different forms and adapt it to our various needs, then we should consider all of the available energy as we tabulate our sources and apply our ingenuity to use as much of it as is convenient.

We now are faced with a number of practical issues, taking into consideration the economics and practicality of using these various sources. Which are most costly? Which are the safest? Which are of the most benefit to society? What are currently the most available, and will that availability run out in the near future? How do we define the near future? What steps should the earth's residents take now to provide adequate world energy tomorrow? Let us now take another look at the energy sources listed previously, describe them, and consider what the energy availability is for each of them.

As we examine each of these sources, the reader should keep in mind that the analysis in section C of chapter 2 shows a present annual energy usage of 100 quads per year in the United States, based on a current global energy usage of 450 quads per year.

B. Solar Energy

Solar energy refers to the energy radiated by the sun and the many forms in which it can affect us. The solar energy that reaches the earth may or may not be intercepted and put to use. If it is not, the energy will heat the ground and/or the atmosphere and eventually radiate out back to space. It is this energy that warms the earth's surface, creates weather, breeds forests, grows food, and supports life.

One characteristic of solar radiation is that it is diffuse. This can be an advantage because it minimizes the problem of having to transport the solar radiation around the earth before being able to use it; on the other hand, it can also be disadvantageous because the lack of concentration as it strikes the earth results in lower temperatures, which therefore limit its application. Another problem is that the energy is available only during daylight and is affected by cloud cover and sun angle. Therefore, it requires storage. Finally, we find that many solar energy applications are capital-intensive, so that the "free" fuel must be balanced against the cost of amortizing a large initial investment.

The sun radiates the equivalent of 12 quadrillion quads per year. This energy radiates into space in all directions, and about half of one billionth of it (less than 6 million quads per year) is intercepted by the earth. If one-hundredth of 1 percent of the intercepted energy could be converted to usable measurable energy (while the rest is reradiated out to space), it would easily satisfy the world's current needs. In the meantime, while on earth it is converted to many forms, used by humankind in various ways, and characterized here in specific types.

It is important to recognize that the term "solar energy" is commonly used to refer only to systems of direct radiation, such as solar panels used on rooftops, providing solar thermal and solar voltaic input. However, in a more general sense, we should understand that almost all the energy sources discussed in this chapter are generated from sunshine and are all products of solar energy. In this book, we discuss each of these energy sources and their formation. Simultaneously, we should be aware that 5.4 million quads of solar energy are annually available to humankind (Figure 1E-1) and that with logical and sensible planning, we can easily capture and use the 450 quads that the world currently uses, as well as the energy growth needed for future generations.

1. Fossil Fuels

In each of the following fossil fuel cases, it should be recognized that these are all the products of original solar energy and are available to be used only once. This can be somewhat troubling to those persons who are concerned about the long-term future, inasmuch as coal, oil, and gas currently account for almost 90 percent of the primary energy consumed in the United States. Of course, if we wish, we can call these energy sources renewable, as long as we're willing to wait over a million years before using them.

Before going on to specific types, let us consider the formation of fossil fuels:

In the development of the earth as we now know it, sedimentation plays an important role. "Sedimentation" refers to tiny mineral or organic particles that were transported by the action of wind, water, or glacial ice and eventually were deposited in the oceans, flood plains, and mouths of some rivers. Most known sediment material is made up of minerals, of neither animal nor vegetable origin. Interspersed in the sediment layers, however, is a small amount of organic material (i.e., carbon-related), which varies from pure carbon to many types of hydrocarbons. This material, formed from living organisms as much as several billion years ago, still contains the solar energy that it received before undergoing sedimentation. Over geological time scales, the organic material became part of a bituminous semi-solid compound called kerogen, which sank within the layers of mother rock under the actions of heat, pressure, and time.

Slowly, the hydrocarbons in the kerogen migrated through the rock layers to reservoirs where they could accumulate as a liquid. Also, as the original hydrogen was lost, sites developed where compounds containing up to 95 percent of pure carbon were aggregated, in solid form. In the deeper sediments, extreme heat produced a pyrolysis in the liquid reservoirs, leading to the formation of large volumes of gas.

The next five sections describe each of the principal fossil fuels. We can conclude intuitively that it is not practical to plan on growing crops that take longer than our lifetimes to reach a useful stage, but fortunately, such crops from the past now exist. But let us not forget that they represent a limited and one-time source that will be depleted during a brief part of humans' time on earth. We happen to be fortunate to be living in the fossil-fuel age but had better switch to something else soon. Among the principal problems associated with the use of fossil fuels are that (1) they contribute to air pollution, producing carbon monoxide and nitrogen oxide; (2) they carry the dangers of possible oil spills or pipeline explosions; (3) they carry a possible threat of global warming via greenhouse gases (e.g., carbon dioxide); and (4) they make the United States economically dependent on other countries, such as the Middle East oil suppliers.

(a) Coal

Coal is a fossil fuel formed from ancient plants that died in peat swamp environments and compacted before they had decayed completely. With continuing heat and pressure, the peat was converted into a number of different forms: lignite (also known as brown coal) is the lowest rank of coal, subsequently developing into sub-bituminous and bituminous coal, all of which are used primarily for steam-power generation; and anthracite, also known as hard coal, which is used primarily for residential and commercial space heating. Anthracite is the coal form with the highest percentage of pure carbon, and it produces the most heat with the least flame and smoke.

As our studies of fossil fuels become more sophisticated, it will be necessary to incorporate complex economic as well as political considerations into the examination of each of the fossil sources as their availability begins to disappear. Here we start with coal, which represents over 80 percent of the fossil energy sources in the world today.

Considering the entire world, US Geological surveys estimate that much of the total coal resource in the ground is not recoverable, because it is either inaccessible or likely to be lost in the mining process. Therefore, the recoverable (economically accessible) coal reserves are estimated to be in the order of 1 trillion tons. The countries richest in coal are the United States, China, Russia, India, Australia, and Canada, with the United States having close to one-quarter of the world's resources.

As shown in Appendix 3, one quad of energy is equivalent to 38.5 million tons of coal. On this basis, the world's economically accessible coal reserve of 1 trillion tons can produce a total energy quantity of 26,000 quads. It is shown in chapter 2 that current world energy use is 450 quads per year, with coal providing 25 percent of this energy. At this consumption rate, the recoverable 26,000 quads of recoverable coal energy would be exhausted in approximately 240 years. This estimate, of course, is critically dependent on the growth of the population and the way that the coal is used. Given that gas and petroleum availability will disappear much earlier, annual coal consumption may well increase and available years decrease (e.g., closer to 150 years).

Coal, the most plentiful of our fossil fuels, is found in beds (seams) that may be as thin as one foot and as deep down as 4,000 feet. In the United States, most of the recoverable reserves are obtained by underground mining. When coal lies near the surface, it can be recovered by stripping away layers of earth surface. This is known as strip mining. Most US low-sulfur coal and medium-sulfur coal is found in the West, whereas high-sulfur coal dominates in the interior and Appalachian regions.

Coal can be used in a number of different ways. The most common way is through pulverizing and burning. Coal also may be gasified or liquefied, in which cases the cost is currently competitive with oil as an energy source. Coal-fired plants are a large source of carbon dioxide (CO_2), although technologies could be undertaken to eliminate such emissions.

World coal consumption is more than 5 billion tons annually, and the United States burns a little over 1 billion tons per year. Close to 90 percent of this is used for the generation of electricity in the United States (this number is approximately 75 percent for the world in general). To produce electricity, the coal is pulverized and then burned, producing heat, which creates steam, which spins turbines. The rotating generators then create electricity.

The national dependence on coal decreased significantly in the middle of the twentieth century, as it fell into disfavor politically, economically, and environmentally. It is a dirty fuel, containing sulfur, ash, and particulate matter in its combustion gases. Deep mining produces health hazards; surface mining can destroy the ecological balance of large regions by soil erosion and acid drainage. When railroads switched to diesel oils, and electric utilities began ordering nuclear-powered generators, coal production dropped sharply, and the cost of operations went up. Air-quality regulations caused a further shift to low-cost substitutes, such as natural gas at regulated prices. The Federal Mine Safety Act of 1969 caused many mines to close, and few new ones have been opened.

In the last few years, however, there have been indications that the trend is beginning to reverse as the squeeze on oil and natural gas continues. We are developing ways to reduce coal pollution through pre-combustion processing and stack-gas cleaning. More dramatically, we are discovering that coal can be physically or chemically converted to forms that are less objectionable to use. For example, it can be changed from its lumpy solid form to a clean liquid or gas that still contains much of its stored solar energy, through processes called liquefaction and gasification. The end products, known commonly as *synthetic fuels*, are really more expensive versions of original coal but are easier and cleaner to use. Coal can also be pulverized and mixed with water to be used as a substitute in normal oil-burning applications. In such cases, the cost savings over oil can be extremely attractive.

Whether used in its original solid form, or liquefied or gasified, or converted to a coal-oil mixture or a coal-water slurry, we should recognize that coal is currently one of the world's major energy resources. But we must also remember that it will disappear eventually, and a replacement plan should be available.

As described previously, the burning of coal can be harmful to the environment. Research is therefore under way to reduce the harmful effects of coal burning by capturing the pollutants or by gasifying and/ or liquefying the original rock form, thus producing energy sources that may be cleaned significantly before being used. This may ultimately limit the harmfulness of coal burning but may be very expensive to implement.

(b) Oil

The word "oil" can have a number of different meanings. When used here, it refers to petroleum or "crude oil," a flammable liquid hydrocarbon that occurs naturally in the rock strata of certain geological formations. When properly distilled, portions of the petroleum are converted into paraffin, fuel oil, kerosene, and gasoline.

Oil, like coal, is formed within permeable rocks by transformation of organic material, from solar energy trapped in ancient plants. The pores within these rocks are filled with a mixture of oil, gas, and saline water. The oil becomes useful when it has concentrated to the point where it can be extracted. The first discoveries of petroleum go back to the ancient Middle East, where this oily substance that seeped through cracks and rock fissures was periodically collected and used in several ways, as a medicine, as a lubricant, or as a source of fire to be used in warfare. However, this knowledge was not transferred to the West and was rarely utilized in the East, and the existence of this substance was forgotten. Then, in the 1850s, a number of citizens in the region of Titusville in northern Pennsylvania became aware of a dark, smelly liquid bubbling up from the rocky ground. By skimming it off the surface of springs, or letting it soak into rags that could be squeezed out, small quantities of this "rock oil" could be collected and examined. The group doing the examination felt that this substance could have a medical application or be used as a lubricant. They soon realized that it could be burned to produce a low-cost highly efficient illuminant that was needed so badly in that period and that was so much easier to use than oil from the sperm whale or a wick dipped into animal grease.

Within the next decade, developers now moving to Titusville recognized that by drilling into the ground and pumping out the liquid, large amounts of "rock oil" could be obtained. The first oil well was drilled by Edwin L. Drake in 1859. This was followed by a market frenzy, starting with several thousand barrels before 1860 and leading to the "industrial revolution" and massive reserves we have today (see chapter 4).

The area near Titusville became known as the Oil Regions. A large number of wells were drilled, and it was soon realized that some would yield oil, but others would not, and that even those that were successful could run dry after a while. The market for lubricant and illuminant (i.e., kerosene) continued to expand, and by 1880 the use of the kerosene lamp was changing American life. In addition to kerosene and lubricant, refineries were now producing gasoline for illuminating buildings and making petroleum jelly (Vaseline) and candles. Geologists became involved to see whether oil sources existed in areas other than Pennsylvania, and then around 1885, oil was discovered in Ohio and Indiana, although Pennsylvania continued to be the dominant production area.

In the late 1870s, Russia discovered that it had an oil source in Baku, and soon after, oil production grew rapidly. In 1901, massive discoveries of oil were made in the United States, in Texas, Oklahoma, and Louisiana. In other parts of the world, oil reserves were being discovered in Sumatra, Borneo, and the East Indies. And then oil reserves were discovered all over the world.

Today, excluding Siberia, roughly 60 percent of all known oil reserves are in the Middle East, with most of the rest split between Africa and the Western Hemisphere. To operate with coordinated action, in the early 1960s a group of countries (Saudi Arabia, Iran, Iraq, Kuwait, Venezuela, and other oil producers later) formed the Organization of Petroleum Exporting Countries (OPEC). OPEC was an influential entity for a while, but it later broke up and lost its influence.

Most people today think of oil when the subject of energy is raised; this is primarily because of the ubiquitous automobile, which uses processed oil (gasoline) to run its engine. Prior to the twentieth century, the main function of oil was to provide light. However, with the advent of the automobile, the internal combustion machine overtook the horse and the coal-powered locomotive, and petroleum took over as the principal element of national power for transportation. Petroleum now represents two-thirds of the oil Americans use every day. It is also used for the manufacturing of various other products.

Today, the oil industry has an enormous impact on the US economy. Seven of the twenty largest corporations in the United States are oil companies. The United States is the largest user of oil in the world, using more than the next six countries combined (Japan, China, Germany, Russia, South Korea, and Brazil). Most of our oil use is in the form of gasoline, used for transportation. A little less than half of our oil is provided from US reserves; the rest is imported from Canada, Saudi Arabia, Mexico, Venezuela, Iraq, and Nigeria.

There are many challenges associated with the continued use of oil. We must find new reserves; we must perfect advanced means of extracting the oil from the earth; we must develop equipment for transporting and processing the crude material; and we must overcome the pollution threats associated with fuel combustion and oil spillage. All of these require massive resources in the form of pipelines, tankers, deep-water ports, and refineries. And yet we began using oil less than 150 years ago and will stop using it within the next hundred years, depending on how fast it is used. Within humans' possible span of a few million years on earth, we live in the brief period of a couple hundred years during which oil is available to us. Yet we have based a massive economic structure and sociological existence on the deliberate burning of this precious and fragile commodity that can also be used as a source of plastics, fertilizer, medicine, and general feedstock.

The world's oil reserves are generally measured in millions of barrels. As shown in Appendix 3, one barrel equals 5.7 million BTUs, and 1 billion barrels equals 5.8 quads of energy. The world's known oil reserves at the present time are estimated to be about 1.2 trillion barrels. About three-fourths of these reserves are in Middle East countries, with almost one-fourth of the world reserves residing in Saudi Arabia alone. Less than 3 percent of the total world oil reserves (about 25 billion barrels) is found in the United States, with an equivalent amount in Mexico. Recent discoveries in Canada have increased its oil reserve to more than 180 billion barrels, which is greater than Iraq's reserve and almost competitive with Saudi Arabia's.

In the United States, oil supplies were interrupted during the 1973 energy crisis, and in 1975 the Energy Policy and Conservation Act (EPCA) was passed, establishing the Strategic Petroleum Reserve, which would maximize long-term protection against oil supply disruptions. This reserve holds a stockpile of more than 700 million barrels of crude oil as an energy supply. Basically, this supply is an instrument that protects the oil market from disruptions such as national strikes or hurricane damage.

There are probably still many natural oil reserves to be found, and it is assumed here that the total world supply could ultimately be as much as two times today's known reserves. However, in the past we have regularly discovered new oil resources at a rate greater than our annual usage. This is no longer true. We are currently at a "topping point," where the annual use exceeds the new discoveries. As this continues, we can expect that the available supply of oil will shrink annually while the cost rises. There is not a great amount of oil in storage, compared to current demands, and it is predicted that there are no longer any great oilfields left to find. Unless we take specific steps very soon to develop and use alternate energy sources, dependence on the disappearing oil will become catastrophic.

On a worldwide basis, total oil usage is approximately equal to current production, at a level of approximately 30 billion barrels per year, or 174 quads of energy. In the United States, current oil consumption is more than 7 billion barrels per year, and more than two-thirds of this is obtained from imported oil. Assuming that oil usage on a worldwide basis remains constant, and that reasonable additional amounts of oil reserves will be discovered, a significant depletion of this energy source will take place in less than the next fifty years.

In the innumerable analyses and discussions of the state of the world oil supply, it has generally been assumed that all automobile and airplane transportation would continue to be dependent on petroleum. Obviously, this is an unwise and dangerous assumption, and we should be devoting a lot of effort to finding other transportation fuel resources, as well as to replacing oil use with alternative sources. This is discussed in several later chapters.

The price per barrel of oil has varied between $10 and $140 in the last few decades and is dependent on many uncontrollable factors. In any case, the world's use of oil will decrease as it starts to run out.

(c) Natural Gas

In addition to the solid (coal) and liquid (petroleum) forms of fossil solar energy locked into the earth by sedimentation, a third massive energy supply exists in the form of natural gas. This is not the petroleum product (gasoline) that we buy at a gas station, nor the propane that we use at a barbecue, but rather an energy-rich hydrocarbon consisting primarily of methane (CH_4), storing the energy remains of tiny animals and organisms that lived hundreds of millions of years ago. The creation of this gas required very high temperatures, and so the gas deposits are generally found more than a mile below the earth's crust. They are generally located in the vicinity of oil deposits but are deeper down. Once discovered, they are brought to the surface by drilling wells and are then refined and brought to the users via pipelines.

For many years, natural gas was considered a waste byproduct and was burned off during oil production. Around 1920, the value of natural gas as a primary fuel with high energy content and clean-burning characteristics was recognized, and its use increased rapidly. In 1958 it displaced coal as the nation's second most important energy source.

Natural gas is colorless and has no odor; when burned, it is an abundant energy supply; and it is clean, emitting very low levels of byproducts into the air. The entire world currently consumes around 80 trillion cubic feet of gas per year, with about one-quarter of that consumption taking place in the United States. About one half of the current use of natural gas goes into industrial and commercial applications, and the other half is split between residential use and electricity generation.

Like oil, the significant exploitation and utilization of natural gas really began in the nineteenth century, and the primary initial use was for illumination. Then with the invention of the Bunsen burner in 1885, it became possible to mix gas and air and create flames that could be used for cooking and heating. Today, with the ease of pipelines for transporting it, natural gas is competitive in price and easy to use in many applications, especially home heating and domestic water heating. We must remember, however, that it is a fossil fuel and will eventually disappear.

Inasmuch as most fossil fuels are buried beneath the earth's surface, it is a considerable challenge to make an accurate estimate of how much of these resources is potentially available but still undiscovered. We are the most comfortable with the resource estimates for coal, but those for undiscovered oil are more difficult to make, so in the previous section it was assumed that the quantity of future new petroleum reserves would be approximately equal in magnitude to those that have already been discovered and are currently being exploited. In the case of natural gas, the problem is more complicated, since these reserves are buried deeper in the earth and represent a large unknown at present. They are usually associated with deep oil deposits and may be several miles below the earth's surface. In such cases, recovery of the natural gas may require deep drilling through a rocky surface.

US natural gas production peaked in 1973 and has been generally flat since then. With decreasing estimates of availability, we can anticipate that from now on the consumption of natural gas will begin to decline. Estimates on the availability of natural gas generally vary, but they all predict reductions. Most of the US resources of natural gas are found in the area of Texas, the Gulf of Mexico, Oklahoma, New

Mexico, and Wyoming. A 2002 estimate by the US Energy Information Administration indicated a US national gas resource of about 1,200 trillion cubic feet (tcf); at the same time, the proved natural gas reserves for the entire earth is estimated to be in the neighborhood of 5,000 tcf.

As before, if we assume that unknown future reserves will be equal to known present ones, we end up with a total natural gas estimate of 10,000 tcf, which corresponds to an energy source of around 10,000 quads. Russia produces slightly more than the United States, and the two countries together provide over 40 percent of total world production. At the current consumption rate of 80 tcf per year, the world's natural gas would have a remaining life of 125 years if world population and energy use stopped growing. Applying conservative and more realistic principles, the life of natural gas is probably between 50 and 100 years. During this period, inasmuch as 1 tcf produces approximately 1 quad of energy, this source can provide the world with approximately 80 quads per year.

(d) Oil Shale and Tar Sands

Shales are sedimentary rocks formed out of ancient clay, with a tendency to split between the thin layers of stratified rock. Depending on the environment during its formation, shale may have a wide variety of contents, such as calcium, alumina, and carbonaceous material. When the shale contains a large amount of organic matter, it has the potential for extracting shale oil and combustible gas for use as fuels.

Again, we are considering a fossil form of solar energy that was originally captured in plants and trapped so that it could not decay completely. Unlike petroleum, which naturally transforms to a liquid, oil shale is retained in a semi-solid form, in the bituminous compound called kerogen. A low-grade liquid oil can be removed by fracturing and processing the kerogen on site or by mining, transporting, and heating the shale, adding hydrogen and using lots of water. The resulting oil can be used for the production of jet fuel as well as gasoline.

There is a great deal of shale oil throughout the world, estimated at more than 2 trillion barrels, and more than half of this resource is found in the United States, with the largest deposits located in the Green River formation in Colorado, Utah, and Wyoming. Much of these deposits are under salt water.

Many projects have been undertaken to extract and use oil shale, and in most cases they have been abandoned as impractical and economically unattractive. In practically every case where countries and large corporations have invested time and money in the development of oil shale, the efforts were found to be too expensive.

Although estimates of oil shale reserves in this country exceed a trillion barrels, there is now no current production activity in the United States. Activity elsewhere has been reduced drastically, although there are still levels of production and research going on in Estonia, China, Brazil, and Australia.

The problems of shale oil recovery are principally logistic, and they are immense. They involve the building of major industrial complexes and support communities, massive rock extraction operations, bringing in great quantities of cooling water over long distances, and construction of transportation networks. The energy required for doing all this and for processing the shale oil is enormous, and the crushed shale will not fit back into the hole from which it was taken. As a result, a byproduct of the operation is the building of mountains.

Despite these severe logistical problems, there is a lot of energy in shale oil, and it must be considered as a potential significant source, with a potential energy availability of a trillion and a half barrels, equivalent to more than 15,000 quads, greater than that for natural gas. Half of this source is located in the United States. The real problem, however, is one that has not been sufficiently emphasized: the net and efficient utilization of energy on a global basis.

Certainly, the large amount of energy available in oil shale is attractive, but is it possible that the (unmeasured) energy required for blasting, transporting, crushing, heating the material, providing and then adding hydrogen, and then disposing of massive amounts of waste is greater than the energy to be extracted? There is a good chance that the answer is yes. The cost of extracting and processing the shale oil appears greater than the value of the energy that is obtained.

Technical documents sometimes refer to bitumen, a hydrocarbon material associated with the formation of petroleum, varying in form from a viscous oil to heavy asphalts and tars. Tar sands, also known as oil sands or bituminous sands, are deposits of bitumens mixed with sand and clay and some water. The extraction process is somewhat similar to that for oil shale, except that the organic fossils have combined with sand instead of with impermeable rock. Here the extraction process consists of adding hot water to the sand and creating a slurry that is then agitated so that the bitumen separates from the fine solids. This bitumen must then be refined to produce a synthetic crude oil. This process requires two tons of tar sand to produce one barrel of crude oil.

Like oil shale, tar sands are found in many countries of the world, with the biggest deposits in Venezuela and in the Athabasca region of Alberta, Canada. In the cold weather, the material is brittle and hard to handle. In the hot weather, it is viscous, sticky, and also hard to handle. Extraction of tar sands also has a serious environmental effect since it requires a large local water source, a lot of energy to boil the water, a large wastewater disposal problem, and potential environmental damage below the surface. When used, it also produces a large addition to atmospheric carbon dioxide.

The economically extractable potential for tar sands is expected to be substantially less than that for oil shale, and some of the same extraction problems prevail, such as wastewater and environmental damage. Therefore, even though the total energy source is in the area of 10,000 quads, it is quite probable that tar sands will never be a justifiable resource.

(e) Synthetic Liquidized Fossil Fuels

The preceding pages describe five types of fossil fuels: coal, oil, natural gas, shale oil, and tar sands. The question then arises as to *how* the energy is transferred when needed, and the answer is usually in the form of heat, by burning these fuels as solids, liquids, or gases. Doing this to solids such as coal can sometimes be inconvenient and/or impractical, and it would generally be desirable, if possible, to convert the fuels to a liquid form that can easily be burned.

There are several techniques for making the conversion to practical liquids. A typical method of coal liquefaction, for example, involves pulverizing the coal to powder form and mixing it with heavy oil or other fluids. Then the temperature is raised to the vicinity of 800 degrees Fahrenheit under high pressure using catalysts, such as hydrogen and iron ore.

The production of such unconventional "oil" becomes of greater interest as the world's natural oil reserves go down. There are large quantities of heavy oil (which has the consistency of molasses) and tar sands worldwide, with major sources in Canada and Venezuela. The United States has about 62 percent of global oil shale deposits, with Russia and Brazil having an additional 24 percent. Again, a significant effort is required to convert these to usable liquid form; this involves extraction, pyrolysis, and hydrogenation. If the production of these synthetic liquid fuels could cost less than the price of crude oil or other substitutes, there could be an attractive supply for hundreds of years.

Unfortunately, the production processes for synthetic liquidized fossil fuels are not easy. The overall process involves a lot of capital, land, manpower, and energy for heat and electricity. As long as the production is economically attractive, the technology will be pursued. Sasol, in South Africa, has been running a coal-to-liquid plant for decades, and new plants are now being built in China, as well as other places around the world.

Interest in a synthetic substitute for petroleum has varied greatly during the last century. The concept of extracting and developing shale oil started in 1925, but it was not pursued very much during the oil glut of the 1930s. Then, during World War II, the army ran out of fuel supplies, so General Patton had the fuel drained out of captured German vehicles, which was an important factor in helping to end the war. In 1944, the United States adopted the Synthetic Liquid Fuels Act, creating a five-year effort to reduce the use of oil by concentrating on coal conversion. By 1950, lots of unforeseen technological problems had arisen. By 1953, the United States started work on better ways to liquefy and gasify coal. Even so, the government generally could not get the price below that of oil at the time, and so the effort was reduced to a low priority.

When a further oil crisis became evident in the late 1970s, Congress authorized an $88 million further investment in synthetic fuel pursuits. When another oil glut appeared in 1986, however, despite the fact that billions of dollars had been spent, this program was abolished.

The development of shale gas is now showing signs of popularity, with new extraction techniques that show promise of expanding the world energy supply significantly. With current technology, there is insufficient permeability in shale to permit reasonable fluid flow, but through drilling horizontally—up to a mile in length—and fracturing the shale in new ways, there is interesting promise in the predicted results. Such technology is being developed earnestly in China but has not yet been undertaken in Europe.

2. Biomass

Biomass is a general term used to describe biological material such as trees, grasses, agricultural crops, and the waste products that these materials may eventually produce. As described in section E of chapter 1, 47 percent of the solar energy intercepted by the earth and its atmosphere is actually absorbed (for a short while) in the earth's surface. Most of this energy is manifested in the form of heat; however, a small portion of it is subjected to a chemical process known as photochemistry. Photochemistry refers to the process in which an unexcited molecule on the earth absorbs a tiny amount (called a quantum) of light energy (a photon) and then becomes a highly excited and energized molecule. It may then break apart, combine with other molecules, or transfer its energy elsewhere.

Appendix 2 provides some brief definitions for understanding how solar energy is captured in living matter, known as plants. This captured energy is the basis of all the fossil fuels that have been discussed in this book. We shall now consider the use of this energy when it is stored in forms immediately extractable (i.e., when it is available within days, months, or even decades) rather than in forms that involve fossil formation use (which takes hundreds of millions of years). This solar energy classification is known as biomass. The portion of the earth's solar energy that is converted into biomass is equivalent to more than five times the total world energy consumption.

Biomass energy therefore refers to energy stored in organic material. In this section we consider some of these principal forms of biomass. As shown in Appendix 2, each of these products contains energy in the form of carbohydrates; we must therefore consider the various techniques by which this stored energy can be converted to forms that will satisfy societal needs. Among these techniques are direct burning (combining with oxygen to produce heat) and decomposition by heating in the absence of oxygen or air, to generate large quantities of gas and/ or fuel oil. The decomposition can be accomplished naturally, with the organic matter permitted to decay in landfills, or artificially, through the heating of the matter in digester tanks without the presence of air or oxygen. The fuels derived from these procedures can be used for generating electric power while simultaneously producing a source of heat. When used as an energy source, biomass releases carbon dioxide in amounts approximately equal to that which was captured in its own growth, and hence no "new" greenhouse gas is created.

Here, we start with wood, the largest biomass energy resource being used today, and then consider other forms of biomass and how they can be used. We know that we can grow a good-sized tree in twenty years and take only a couple of days or a couple of hours to extract its energy by burning. Then it is gone. We should therefore consider whether there is a more efficient way to capture the photosynthesized solar energy. For example, kelp grows as much as a foot a day. Alfalfa too is a fast grower. Weeds spring up rapidly. Suppose we take the solar-intercepted area equivalent to that of a tree, and instead of growing a tree, we harvest several crops a year of fast-growing grasses and plants. It is likely that the collected dried plants after twenty years will produce more BTUs than the tree would have. We have always thought of growing crops for their food value. Perhaps we should think of growing them for their heating value. Several such examples of biomass are discussed in the sections to follow.

In considering the potential for biomass, we should first be aware of the total available land acreage in the United States. The forty-eight contiguous states have a land area of more than 1.9 billion acres, and Alaska and Hawaii have a few hundred million more. The breakdown of this land area (rounded off in most cases and subject to continuous gradual change) is shown here, broken into a number of general categories:

Land type	Area (in acres)
Rural residential	80 million
Urban	60 million
Cropland	
corn	80 million
soybeans	80 million
alfalfa hay	60 million
wheat	60 million
other crops	70 million
currently idle	60 million
Range and pasture grassland	700 million
Timberland (virgin forest)	500 million
Grazed forest	150 million
Parks, wildlife, industrial, defense	300 million
Other miscellaneous	<u>150 million</u>
Total US land area:	2,350 million

More than 60 percent of the 2.35 billion acres of total land area in the United States is privately owned; the rest is owned principally by the federal and local governments.

As society becomes aware of the significant energy potential that is available from the sources contained within forests, agricultural lands, crops, and waste facilities, the development and usage of these continuing facilities will become a major portion of the solution to the problem of the ultimate disappearance of fossil fuels. Some effort must obviously be devoted to establishing a safe and sustainable supply system. This involves the tasks of collection, storage, and delivery of agricultural residues, including processing techniques. Animal manure is also a potential biomass resource.

Using biomass, rather than fossil fuels, is an enticing proposition, even though the energy comes from the same (solar) source and is captured by plants in the same way. In the case of fossil fuels such as coal and oil, the material is heated and compressed during the sedimentation process, and sulfur, mercury, and other noxious elements are added to the fuel and produce serious effects when the fuel is later burned. On the other hand, biomass does not contain these dangerous elements and will not produce dangerous emissions when burned.

At the present time, across the United States there are approximately 100 operating biomass power plants. Most of these are in California, but many other states are now getting involved, particularly for generating local electrical power.

The following sections cover three categories that are classified here as biomass. The first, identified here as "wood," refers generally to trees and to the forests that we do not propose to destroy. The second, called "organic wastes," describes all those biomass sources that can be converted to usable energy through burning. This section includes grasses, which grow to maturity on an annual basis. Finally, we describe "biofuels," liquid energy sources such as ethanol based on various plant products including switchgrass. At the present time, the total use of biomass resources enables this country to produce about 3 percent of its total current requirement of 100 quads (from section C of chapter 2). As later discussed in chapter 5, this amount of 3 percent can be increased substantially.

(a) Wood

The hard fibrous substance that makes up the greater part of the stems and branches of shrubs and trees is called wood, and it can continue to store the chemical energy for many years until this energy is recovered through oxidation, which may be slow (decay) or fast (burning). From

49

humans' "taming" of fire 150,000 years ago until the nineteenth century, wood was the primary energy source used by humans. Even today, half the world cooks with wood, and far more than three-fourths of the energy in most of the undeveloped countries comes from wood. In the United States, however, no effort has been made to manage forests and develop advanced harvesting and processing techniques to develop wood as a major energy source. Despite this, many homes in this country use wood as an alternate source of heat, in fireplaces, stoves, and furnaces. Home fireplaces can be attractive and pleasant, besides providing warmth through radiation and convection.

If we consider the purpose of a tree to be solely for the collection of solar energy over a 25-year period, we should now consider whether there are better ways of using the area represented by each single tree over this long time period for energy collection. For example, eucalyptus trees grow more rapidly than oak; perhaps four eucalyptus trees could use the same circular area and grow to maturity in eight years. This way we could accumulate the wood of twelve eucalyptus trees in the same time that we grow one oak tree. In undeveloped countries, wood from trees will continue to be a major energy source for heating and cooking. However, as shown herein, the United States can generate a significant amount of energy from forest products without destroying the forests themselves, which play a major role in cleaning the air and regulating our atmosphere.

Although not generally considered a fossil fuel, wood can be fairly old, except that the time scale is one of decades rather than millions of years. In this book we do not consider a program for growing new forests in the world, but rather consider any new wood growth and shrubs in the category of "organic wastes." Neither do we consider the wood that has died and changed its form into coal or oil, since these energy forms have been discussed separately. Instead, we recognize the trees that exist today in the form of forests, most of which are unused and untapped, which cover almost 30 percent of the earth's land surface.

There are no accurate figures on the total size of the world's forests in terms of volume or mass of all the timber therein. It is reasonable to estimate, however, that the total stock is about forty times the average growth (average forest tree age equals forty years) and that the timber depletion rate is in the order of the new growth rate (although in some specific areas there is much concern among environmentalists that the forests are being denuded too rapidly). Using such assumptions, it can be estimated that the total growing stock in the forests throughout the world is over 4 trillion cubic feet of lumber, which corresponds to a total energy capacity of 2,000 quads.

No one is going to propose that we cut down all the forests. They represent a critical element in the earth's ecological structure. Perhaps wisely or unwisely, some portion of this energy resource could be used. We have been willing to use oil with impunity, and it is no less vulnerable a resource. We should recognize that thinning of forests can improve their health and reduce the threat of serious fires. On the other hand, maintaining large shade trees can improve wildlife habitat, while simultaneously enabling other forest vegetation to generate needed power. For the present, let's note the existence and capacity of this energy supply.

Considering all the other available sources, it is not suggested here that a program be undertaken in the United States to grow trees for the purpose of providing wood as an energy source. As is shown in the next section, wood does provide a significant source, but generally as an organic "waste." This can be in the form of old furniture, broken pallets and trusses, trimmings, and other wood wastes. It takes a conscious and continuing effort to maintain a satisfactory environment with healthy forests, and it is not recommended here that we ever take actions that could destroy the forests or the environment.

(b) Organic "Wastes"

In general, municipal trash collection systems make an effort to separate glass, plastics, and paper, all of which are targeted for recycling. In addition, we must recognize that such trash also contains inorganic and sometimes toxic material such as metals, batteries, and possibly hazardous products. An effort should be made to separate these toxic items, with the remaining trash sent on to a garbage dump, despite the fact that it contains a wealth of biomass, such as dead branches, yard clippings, sawdust, decayed or unused food, wood waste, old paper goods, and rags. "Urban wood wastes" include furniture and broken pallets. To this we can add dedicated energy crops and trees, agricultural feed crops, and aquatic plants. In California, the first small biomass plants began producing electricity in 1982, and today the California Biomass Energy Alliance (CBEA) represents thirty-seven biomass-fueled power plants throughout the state. Starting with sawmill residue, these plants have expanded to use forest thinnings, agricultural byproducts, and urban wood waste. California currently consumes 7 million tons of organic waste per year, about 25 percent of all the waste deposits in the state. This activity is a considerable help to forest management, since many forests are choked with trees, and thinning them protects the large trees and reduces the possibility of fires.

In the United States today, utility power plants using wood waste as a fuel provide a total of 8,000 megawatts, with individual plants producing up to 80 megawatts. This is a pretty small number when compared to our current electricity requirements.

This "trash" can be used as an energy source in several different ways. It can be fed into furnaces and burned, producing heat that boils water and is converted to electricity. Or it can be collected in landfills, where it will decompose and give off large quantities of methane, which can then be piped to electricity generation plants. One concept of generating methane from the decomposition of animal manure has been suggested and tried, but the results so far are not attractive. The costs for collecting and processing the manure are high, significant pollutants are produced, the disposal problem is difficult, the benefit of manure for producing soil fertility is lost, and the energy that is generated is not worth the expense. Very little investment has been made in this energy sector, and no interest has been shown in generating methane from animal manure. However, use of organic trash without concentrating on manure as a primary source is still a very attractive and potentially economical concept.

Unlike wood, grasses grow to maturity very rapidly and adapt to many areas and environments. There are many thousands of species of grasses, which are the principal plants in large areas throughout the world. These are generally annual or perennial plants, with fibrous leaves and stems much softer than are found in "woody" plants. The dry seed-like fruits of grasses are called *grains*. Cereal grasses, such as barley, wheat, oats, rice, corn, and rye, provide grains that are a major food for much of the world, whereas other grasses can be grazed by domestic and wild animals. In addition to hay, fescue, sugarcane, sorghum, and a wide variety of lawn grasses, there are those that are processed to produce liquor, molasses, corn starch, alcohol, and paper.

Switchgrass is one of the dominant North American grasses, found in prairies such as the Great Plains, many pastures, and gardens. It is perennial and hardy, grows to heights of six to twelve feet, and survives well in heat, drought, and flood. It begins its growth in the spring and is used for livestock feed as well as ground cover. Switchgrass in the southern United States produces about eight tons per acre, which is more than three times the yield obtained from normal hay. Its energy content is comparable to that of wood, with much lower initial moisture content. When used for energy production, it can be cut twice a year. Other attractive features of switchgrass are that its extreme network of stems and roots below ground enable it to prevent normal erosion, and it removes significant carbon dioxide (CO_2) from the air as it grows. Compared to corn, for example, growing and using switchgrass requires far less energy, without extensive need for tractors, farm transportation equipment, and harvesting efforts.

In addition to being an easily grown and efficient source of energy that can be obtained by direct burning, switchgrass is also an attractive candidate for the production of biofuel, such as ethanol. This is discussed in the next section.

The Energy Efficiency and Renewable Energy (EERE) center of the US Department of Energy recently reported that 510 million tons of biomass could be available at less than \$50/dry ton and would be equivalent to eight quads of primary energy. The challenge is then one of collecting, storing, delivering, and processing the agricultural residues. The following paragraphs describe the principal categories of organic wastes that make up the 510 million available tons. This information is based on a 1999 study conducted by Oak Ridge National Laboratory, University of Tennessee, and Science Application International Corporation, considering five principal categories of biomass feedstocks:

1. Forests in the United States cover more than a half billion acres; they are used to produce annual supplies of logs, papers, veneers, and other products. Because of the extensive harvesting within the forests, there are large quantities of wood residues that can be used as a biomass source, including partially decayed wood, logging residues, and excess saplings. To make use of this residue, we must consider the costs of collection, harvesting, chipping, loading, hauling, and a return for profit and risk. The current estimated quantity of US forest residues available for energy conversion at a reasonable price is 45 million tons.

2. Wood mills are the factories where raw wood is brought and converted to lumber, pulp, paper, board, and fiber products. In the process, the residues that are created consist of bark, slabs, shavings, and sawdust. As before, the residue available for conversion to energy is dependent on the price at which it can be sold. Much of this cost goes

into transportation, so it is important to keep the energy production facility reasonably close to the mills (e.g., less than fifty miles). This category provides 90 million tons of convertible waste.

3. It was shown earlier that approximately 350 million acres of land area are used for growing crops. Those crops that have the most attractive potential for energy conversion are corn stover (stalks without the ears) and wheat straw. In addition, some energy feedstock is available from barley, oats, rice, and rye. These currently can be sold by farmers to companies that produce insulation, paper, chemicals, and other products, so the costs for collecting and transporting become very important in assessing the quantity that might be available for energy conversion. This quantity is currently estimated at about 150 million tons.

4. Urban wood wastes include yard trimmings, site clearing wastes, wood packaging, pallets, and other wastes that are generally disposed of in landfills. The estimates are rough, and they vary across different parts of the country.

5. Instead of, or in addition to, collecting agricultural residue, it is possible to plant high-energy rapid-growing crops such as hybrid poplar, hybrid willow, and switchgrass. A model analysis performed by the Policy Analysis Center at the University of Tennessee, working with the Department of Energy, takes into account the planting density, the reharvest and replanting times, and all associated costs. The analysis concludes that it is not feasible to undertake such a program at a cost of $30 per dry ton of crop output, but at $50 per dry ton, the annual output could be reasonably attractive.

Applying the best estimates that can be made to analyze the costs of each of the five processes just listed, and adding them together, the following table shows the conclusions that have been drawn with respect to US organic waste energy availability:

Delivered price	Quantity (dry tons/yr)	Quads/year*
<$20/dry ton	20 million	0.3
<$30/dry ton	105 million	1.7
<$40/dry ton	315 million	5.2
<$50/dry ton	510 million	8.5

(c) Biofuels (e.g., Ethanol)

Some of the energy sources that we have already considered in this section (e.g., coal), and some that we will be considering (e.g., wind, nuclear fission) are of value in serving many of humankind's needs without going through a conversion to liquid form. However, much of today's energy use, particularly with respect to automobile and airplane transportation, has been dependent on fluids and gases, and it is therefore convenient to have available an energy source that is in fluid form (such as oil). We have just seen that biomass can be a significant energy source, but it is transformed from solar radiation directly to plants and grasses when stored on earth. Therefore, we see that it can be very valuable if we can convert the biomass energy into fluid and gas forms that can then be stored and used easily in the transference to kinetic energy. Liquid and gaseous energy sources made from biomass are known as biofuels.

Biofuel is attractive because it is a renewable energy source. Some agricultural products grown specifically for liquid biofuel use are corn, soybeans, flaxseed, sugarcane, and palm oil. Other sources, converted to gas use, include straw, timber, rice husks, switchgrass, and biodegradable waste.

A very important energy source, known as *ethanol*, may be obtained from wood, from organic wastes, or from agricultural products. Ethanol, commonly known as "grain alcohol," "ethyl alcohol," or just plain "alcohol," is a flammable, colorless, tasteless chemical compound made of carbon, hydrogen, and oxygen (C_2H_5OH, or C_2H_6O), which can be used in intoxicating beverages or as a solvent. The process of producing ethanol is described in Appendix 4. With today's technology, almost three gallons of ethanol can be produced from a single bushel of corn.

Besides use in alcoholic drinks and tonics, ethanol can be used as a solvent, cologne, and sanitizer and in perfumes and paints. For energy considerations, the most attractive use of ethanol is as a rocket fuel, an automobile motor fuel, or a motor fuel additive. In this section we first consider how ethanol may be used as an energy source replacement for oil, and then we consider the potential amount of energy from ethanol that is available.

At the present time, the fuel ethanol industry in the United States includes more than a hundred ethanol plants, producing more than 5 billion barrels annually. This is most commonly used for "gasohol," which is standard gasoline mixed with 10 percent ethanol and which is widely sold in corn-grazing regions such as the Midwest. Most automobiles in the United States today are powered by straight unleaded gasoline. If ethanol is added to the gasoline to represent 10 percent of the mixture, no engine modifications are required, and the fuel mixture, officially called E10, is cleaner-burning than straight gasoline. This implies that if sufficient ethanol were available and put into use as E10 today, there could be an immediate reduction in the total gasoline supply by 10 percent. Of further interest, however, is the present research and development of flexible fuel vehicles (FFVs). These are vehicles that not only can run on standard unleaded gasoline, but also can use fuels with a biofuel content of up to 85 percent (called E85 fuel). Most of the new cars sold in Brazil are FFVs, and the Brazilian production of ethanol today exceeds four 4 gallons. Besides Brazil and the United States, several countries that are now using gasohol include Canada, Thailand, India, China, and Japan.

At the present time, FFVs are being developed by a number of manufacturers, including Daimler Chrysler, Ford, General Motors, and Nissan. If the car is not an FFV, it may still perform well at ethanol percentages above 10 percent, but may not yet be covered by warranties. In the meantime, the use of E85 fuel is increasing continuously throughout the country.

Up to the present time, most ethanol has been made from corn, although it can also be produced from sugarcane, wheat, oats, barley, potatoes, sorghum, and a large variety of grasses. Switchgrass shows one of the most attractive potentials. There are now well over 100 ethanol plants in operation in the United States, and the number is growing dramatically. Within the next decade and with continued commercialization, we can generally expect that at least 20 percent of current gasoline usage will be replaced by ethanol. In this country, this would be equivalent to several quads per year.

Aside from ethanol, another attractive biofuel with lots of promise is *biodiesel*. This is manufactured from vegetable oil made from such crops as corn, canola, or soybeans. Biodiesel is a clean-burning and renewable fuel that could easily power cars, trucks, trains and boats. It could also be economically attractive to farmers to produce seed crops that can be converted to vegetable oil. One of the problems with biodiesel is that it gels easily. Therefore, although it is an attractive energy source inside a building or in warm weather, it tends to gel when the temperature gets below freezing.

3. Direct Solar Radiation

All the foregoing sources of energy have been based on the use of photosynthesized solar energy. There are other forms of solar energy that we might now consider. The most prevalent of these is *direct solar radiation*, which can be exploited in a number of different ways.

As we have seen, solar radiation heats the air, the ground, and the oceans and is reradiated into space. While on earth, it is this energy that provides the temperature gradients and thermal environments that sustain life on earth. If we choose, however, we can intercept much more of the radiation and convert it to other forms to suit our needs and desires. For example, if all the radiation falling on the roof of a single-family home could be captured, it could supply all of the needs of the household. Solar energy is already used for cooking and water heating throughout the world. If collected over large areas and concentrated, it can produce very high temperatures to run boilers and furnaces. For space applications, it is used to radiate photovoltaic cells, producing electricity directly.

One argument sometimes voiced against the exploitation of solar energy is that it is diffuse, and the task of harnessing it is capital-intensive. Also, because of its intermittent nature, techniques must be developed for storing the energy when the sun is not shining. And so, although the problems of energy transportation are minimized, we do have to consider means of concentration and means of storage. There are a number of practical techniques that can be considered for such concentration and/or storage, but the selection of the most economical is yet to be determined.

Solar radiation can be intercepted in many ways. We can use flat collector plates, photovoltaic cells, or curved-mirror collectors. We can design structures to take passive advantage of solar radiation characteristics. We can take advantage of the thermal gradients produced by solar energy absorption in the oceans, to run generators; we can make these into useful energy sources by using large areas of water so that floating islands or large tankers are required. The energy can then be extracted from the ocean and shipped to nearby land in the form of hydrogen, ammonia, or electricity. We can also use the evaporation phenomenon to build "solar stills" that make fresh water out of salt water. The applications for the exploitation of solar radiation are many. Some of the principal ones are covered as this discussion continues.

As we saw in the energy balance discussion of chapter 1, the earth and its atmosphere receive 5.4 million quads per year from the sun. We have also seen (in chapter 2) that humanity requires about 450 quads per year to survive comfortably on earth, and of this amount the United States uses about 100 quads. Compared to the total solar energy reaching this earth, these requirements appear to be very small. The distribution of the total solar energy over the earth's surface varies with seasons, latitudes, and atmospheric conditions, but in any case it is far greater than humans' anticipated usage. Yet as we have seen in each of the previous sections, diverting each form of available energy to serve our needs is a struggle, and the techniques described in this chapter so far are all dependent on the use of solar energy *after* it has been absorbed in the earth. Let us now consider using solar energy at the same time that it strikes the earth. This leads to a basic question: how much solar energy can we capture and use effectively?

The direct energy sources that we now consider are solar thermal, solar concentration, and solar voltaic.

(a) Solar Thermal

When we feel cold and the sun is shining, we can go out into the sunshine and expose our skin to the solar radiation, where the infrared portion of the spectrum is felt in the form of heat, and we may feel comfortable. This is not a great idea, however, since other portions of the radiation (e.g., ultraviolet) can be damaging to the human body. It is wiser to use the sunlight to heat walls, air, and water needed to produce comfortable environments.

The simplest, as well as the oldest, use of solar energy is use of *passive systems*, one of which merely involves pointing the buildings in the right direction. Early American Indian cliff dwellings were positioned to catch the sun's rays in the winter, while being protected by shadows in the summer. In today's buildings, we can use south-facing windows and skylights to maximize the capture of heat as well as light. This can be combined with overhangs that provide shade when the sun is high in the summer. A greenhouse is a common and good example of passive solar heating.

One of the efficient ways to make use of the sun is by incorporation of solar water-heating systems. The most common of such systems involves the use of *flat-plate* collectors. This is typically an insulated metal box that may be several feet in width and length, but only a few inches high. The top is glass or plastic, to prevent heat from escaping, and below that is a series of copper pipes (flow tubes) that wind back and forth, carrying water from an inlet at one end to an outlet on the other. The collector sits on the roof, and depending on the design, the water may flow sequentially from one collector to the next. Below the tubes, there is generally a dark-colored absorber plate, with a layer of insulation below that. The tubes are usually painted black to maximize heat absorption. The potable household water traveling through the tubes is heated to temperatures approaching 200 degrees Fahrenheit. Such systems can be used to provide household hot-water needs, and they can also be simplified in design and used for pool heating. According to the US Department of Energy, hot water constitutes about 14 percent of our annual fuel bill.

The preceding descriptions refer to passive systems that are normally used in regions that are rarely subject to freezing temperatures. Such systems rely on gravity and the tendency for water to circulate naturally as it is heated. In an active system, a thermometer senses when the solar collector is ready to heat water, and then the water is pumped through the collectors and stored in a conventional hot-water tank. In areas where it can get very cold, the solar thermal systems may add antifreeze solutions to the water and then pump it into heat-transfer units where it warms the household water in the hot-water tank, but the two liquids never mix. Such systems are called indirect. They heat a fluid that heats the water you use. They are often called closed-loop systems.

A different design of flat-plate collectors that can be used for solar space heating does not require tubes in the construction. Instead, cool air enters one end of the collector and then flows through or along absorber plates that collect the solar radiation. The heated air may then move to subsequent plates through the use of a fan that is part of the air circulation system.

There are many sophisticated design modifications of these basic flat-plate collectors, and the frequency of their use is increasing; more advanced systems can bring the water to much higher temperatures, to meet various industrial requirements.

To heat swimming pools, the system can be significantly simpler, since there is no requirement to go from the heated water to the inside of the building. In fact, merely painting the bottom of the pool with a black heat-absorbent coating can eliminate the need for a separate furnace on sunny days.

Another attractive use of solar thermal energy is based on the development of *solar ponds*. The concept of solar ponds has received much attention and is currently being used in a number of installations in Israel. Here the system uses a thermal gradient across a depth of water, but instead of ocean depths, the ponds may be only a few feet deep. Minerals and/or salts are introduced into the pond varying from very low concentrations at the top to very high concentrations at the bottom. As a result, the solar energy is collected at the bottom, producing a reverse temperature gradient in that the bottom of the solar pond may be hundreds of degrees warmer than the top surface. The extreme temperature gradient over a short distance can lead to efficient thermal engines for energy production.

The market potential for the use of this solar thermal energy is massive when we consider practicality, availability, and cost.

When we calculate the amount of energy that is required for producing heat, we do not refer to applications where sufficient heat is already there. For example, in most equatorial regions there is sufficient solar warmth to keep people and houses comfortable without need to install home heating systems or wear heavy clothing. Recognizing this, we should now consider whether the same solar warmth can be efficiently utilized in temperate regions, or even polar areas, where the style of life and natural weather conditions are insufficient to provide the level of warmth that humans seek.

Well-designed solar thermal systems can be relatively attractive and often resemble sleek skylights. Depending on the utilization and geographic location of the single-family house, an efficient solar thermal system can be installed for a total price range of $2,500 to $4,000. The resultant annual savings in gas and/or electric bills will generally recover this investment in less than a ten-year period, after which the savings continue indefinitely, while the demand on the utilities has been decreased.

In making estimates of potential quad production using direct solar energy, let us first consider the simplest form of direct solar heating of homes—that is, using flat-plate thermal collectors on the roofs of single-unit homes. These homes may be attached or detached and are generally one-story (neglecting mobile homes for the present). Studies recently performed for the US Census Bureau report a total of about 120 million existing homes in this country. About 75 million of them are candidates for direct solar heating through the use of flat-plate roof collectors. For large commercial buildings and other structures, the collectors may be ground-mounted, with the heated fluid pumped to large storage tanks and used as needed. Such sophisticated systems may utilize pumps, reservoirs, and controllers to regulate the fluid flow.

We have seen that there are many different units for expressing and measuring energy. In the case of home heating, which is generally done by gas, charges are commonly based on the cubic feet of gas used. The number of cubic feet is converted to therms in the bills that are sent to customers. On the other hand, measurements of solar energy exposures are often expressed in terms of kilowatt-hours (kWh). As seen in Appendix 3, one therm is equal to 100,000 BTUs, and one kWh equals 3,412 BTUs. To be consistent in this analysis, we will convert the therms and kilowatt-hours to BTUs and eventually to quads, as described in Appendix 3 of this book.

Suppose the owners of those 75 million single-family homes decide to convert to rooftop flat-plate thermal collectors. The immediate energy savings would be immense: about three quads.

We get that figure by estimating that in the continental United States, each square foot of flat surface exposed to direct sunlight receives a little over 90 watts at peak, varying with the angle of the sun and the geographic location. Assuming an average of six productive hours of solar energy in a day, the energy absorption is then 6 × 90 = 540 watt-hours per square foot per day. If we then install an average of 60 square feet of solar panels on each rooftop, the total solar direct thermal energy received in all the homes is determined by the following equation:

(540 watt-hours) × (1/1000) × (3412 BTUs) × (60 sq. ft.) × (365 days/year) × (75 million houses) = 3,030,000,000,000,000 BTUs per year = 3.03 quads per year

Since most of these assumptions are approximate, it is safe to conclude that the annual US energy saving from this process can be in the order of three quads.

(b) Solar Concentration

Up to this point, we have discussed a number of available solar sources that are all relatively limited in size when we recognize that the world needs a minimum of 450 quads annually for the indefinite number of centuries to come. We also recognize that fossil fuels are a form of solar energy received by the earth a long time ago. Biomass sources were received more recently but are also limited forms of solar energy. Other solar thermal systems include wind and hydropower. Combining these observations with the portrayal of solar energy illustrated in Figure 1E-1, we recognize that only an extremely tiny portion of the earth's incident solar energy is used by humans. We have just seen that if we can intercept some of the solar energy that falls on flat plates on our rooftops, we can use it to provide significant warmth. We also see in the next section of this chapter that with a specially designed flat plate, we can convert the incident solar energy directly to electricity.

However, these references apply to the use of incident solar energy that happens to strike a plant, a tree, or a flat plate on a rooftop. What about the direct solar radiation that falls on unused land, oceans, deserts, backyards, mountains, and empty fields? The fact is that more than 99.99 percent of the solar energy that strikes the earth is permitted to escape without being intercepted or converted to any useful form.

Recognizing that fossil fuels, currently providing 90 percent of all world energy use, will be fading away, it is logical to establish a plan for making use of some of the solar energy that is currently not being intercepted. To do this, we must collect the radiant energy over large areas and transport it into facilities where we can convert it into electricity and other usable forms. The facilities for doing this are often called concentrated solar power (CSP) plants.

In the solar thermal applications just described, the amount of energy collected is dependent on the area that the collector plates can accommodate, and this is normally limited by the size of the house and its roofs and the availability of yard space. For industrial applications, however, where the locations of factories and facilities are variable, consideration can be given to large areas of sunlight as part of the energy collection process. This energy can then be collected and concentrated through use of massive parabolic dish reflectors to produce narrow high-energy beams that are focused on the energy receiver. Here the energy is collected in a heat-transfer fluid, such as oil or water, which then drives a conventional turbine to produce electricity. The more ground space that is available for mounting parabolic solar reflectors, the more electrical energy that can be generated and stored.

The concept of rooftop solar panels is relatively straightforward, but it is limited by the number of square feet of flat solar thermal panels that can be practically installed on the sun-facing portion of homeowners' roofs. We have conservatively estimated that this could provide three quads of US energy needs. We now consider a more sophisticated means of providing solar energy for heat: by designing collectors that are not roof-mounted and that intercept huge solar areas for reflection and concentration. There are a number of common ways of concentrating energy from the sun. Three of these that are in use today are troughs, dishes, and towers.

Troughs. These linear concentrators are currently used in a number of places in the United States, each presently generating up to 80 megawatts of electricity. New ones being developed are expected to each generate over 250 megawatts.

Troughs are long parallel rows of parabolic mirrors that concentrate the solar radiation onto a tube running the length of the trough. The tube is filled with a heat-transfer fluid, such as oil, which absorbs the heat until the temperature approaches 1,000 degrees Fahrenheit; it is then stored or used in a heat engine to produce electricity. As the sun moves during the day, the troughs can tilt so that the focused radiation on the tube is maximized.

Dishes. These are parabolic collectors similar in shape to those used for satellite television, but much larger, as much as a hundred times more. All the sunlight that strikes the disk is focused onto a single receiver, where the energy can again be stored or converted to electricity. As with the trough, the dish is rotatable and is programmed to point at the moving sun.

Power towers. In some locations (where a lot of land is available), the single disk is replaced by an array of many flat movable mirrors, with each mirror reflecting its radiation to the same focal point, producing working fluid temperatures of more than 2,500 degrees Fahrenheit. A version of this design is commonly known as a power tower.

These three types of solar concentration systems are too big to consider for mounting on roofs, but the only limit on size is how much land is available. In considering this, we may first review how land area relates to energy availability. This is summarized in Table 3B-7. As shown in this table, when exposed to direct sunlight, one square yard receives about 833 watts of solar power. Assuming six productive hours of solar energy per day, the amount received in one day will be 5 kilowatt-hours per square yard.

System efficiencies can be close to 100 percent if we use the solar energy in the form of heat. However, if we are going to convert it to electricity, the efficiency can drop significantly. For example, photovoltaic systems, described shortly, are only 15 percent efficient, with the rest of the incident energy being lost. In the cases of the solar concentration systems described here, it is reasonable to assume system efficiencies of 25 percent, as used in Table 3B-7.

The exposed surface area used in this table varies from 1 square yard to 10,000 square miles. Let us now consider what kinds of areas are available for use in the United States. Note that the concentrated collectors do not require very flat land. They can be built, if necessary, on hills, sides of mountains, lake beds, sandy plains, basins, plateaus, and sedimentary isles.

Table 3B-7
Energy Availability in Sunbelt Regions of the United States

Surface area exposed to direct solar radiation	Power in watts (W) or kilowatts (kW)	Energy rec'd in 6-hr day	Energy/day available as electricity (25% efficiency)	Available as energy inquads (Q)(1Q = 293B kWh)
1 sq. meter	1 kW	6 kWh	1.5 kWh	
1 sq. yard	.833 kW	5 kWh	1.25 kWh	
1 acre	4,032 kW	24,200 kWh	6,000 kWh	
1 sq. mile (or 640 acres)				.0048 Q/yr
10,000 sq. miles (e.g., 100 × 100)				48 Q/yr

First, consider the availability of desert land. Deserts in the United States are quite small when compared to Saudi Arabia or the Sahara, which comprise 3.5 million square miles of desert. However, we do have 200,000 square miles in the Great Basin (Utah, Nevada, Oregon, Idaho); 54,000 square miles in the Mojave (Arizona, Colorado, Nevada, Utah, California); 120,000 square miles in the Colorado plateau; 175,000 square miles in Chihuahua; and 120,000 square miles in the Sonoran desert (Arizona, Colorado, Mexico). Note that for the present we are ignoring millions of square miles of available land in the Arctic.

Let us assume that 3 percent of the total land of the five previously listed deserts is devoted to the concentration of solar energy. This corresponds to 20,000 square miles, which is seen in the table to produce approximately ninety-six quads per year, providing almost all current US energy needs. In addition to the primary function of solar energy collection, some additional area would be required for administrative functions, operation buildings and facilities, paths for traveling through the areas, and so on, but all these areas will probably not constitute more than an additional thousand square miles.

The use of 20,000 square miles to produce almost all the country's energy needs is not an extreme or unrealistic concept when we recognize that during World War II the American army used more than five times this area as training facilities for our soldiers and that the total US land area is more than 3.6 million square miles.

In considering the installation of solar concentration systems, not much effort has been devoted to the determination of the extensive water use that may be involved. Deserts are attractive for capturing solar energy, but it must be recognized that they are, by definition, dry. A significant amount of water is required for cooling the plants, washing the mirrors and panels, and enabling the facilities to function. This can represent many millions of gallons per year, and this factor must be included in any development plans for collection of significant amounts of solar radiation energy.

(c) Solar Voltaic

A *photovoltaic* (PV) cell, also known as a solar cell, is made of a semiconductor material such as silicon and can convert sunlight directly to electricity. These devices were first used in sophisticated space applications to provide power to orbiting satellites and spacecraft. As their manufacturing costs have gone down, they have become useful for powering highway lights, emergency telephones, and calculators and communication devices. Their use is currently dependent on the availability of refined silicon. Nevertheless, the worldwide production rates are still projected to grow significantly, since they are economical after the first few years following initial installation.

The two previous sections discuss direct solar energy that can be used as soon as it is absorbed. In the "Solar Thermal" section, we saw how energy is immediately converted to heat and is used to provide direct warming to the household. In the section "Solar Concentration," we saw how energy is captured from large areas and can then be stored and converted into electrical energy when needed. In this section, we consider the concept of converting incident solar energy directly into electricity, without going through the storage of heat.

The materials initially making up solar cells have been various forms of silicon or thin-film deposits of semiconductors, connected and mounted in a frame with a glass cover. When the sun's waves hit the PV cell, electrons are excited and move back and forth, creating electricity. Since PV cells do not require batteries or power lines, they can be used to provide power for remote locations, space vehicles, ocean vessels, emergency roadside telephones, and island communities. They can be mounted on rooftops and can be very convenient for reducing costs at times of peak demand. The electricity they generate is direct current (DC); for convenient use in most US homes, an inverter can be installed to convert the electricity to alternating current (AC). When the photovoltaic cells are arranged on a flat panel and connected in series, they can produce many watts (or even kilowatts) of electricity while exposed to the sun. The earliest applications of PV solar cells were for satellites and spacecraft, many of which are currently used in the telecommunications industry.

When used in homes that already have electricity connections, the solar array can be tied into the electrical grid, so that the charge to the homeowner is only for the net electrical usage after the solar panel electricity is subtracted.

Most of the current worldwide usage of photovoltaic systems takes place in Germany, Japan, and the United States. Solar voltaic systems are quite expensive. Because of increased demand, the price of crystalline silicon cells has increased, and the availability is much less than it was originally. Much research is currently under way to find other material that will capture energy from sunlight efficiently and at lower cost than silicon. Later in this book we consider the economic effects of intensive use of PV energy systems.

The concept of installing a system that can convert solar energy directly to electricity is very attractive, but the installation cost can be discouraging. Of course, the fuel—sunlight—is free, abundant, clear, and quiet. However, the initial cost is substantial. It involves basic installation, the cost of an inverter to switch from direct to alternating current, and connection to a main service panel. In general, the cost of this installation is recovered within five years. The electricity provided by the PV system is then free. The system may be connected to the utility-provided grid, in which case the homeowner may be credited by the utility company for electricity supplied from the home to the grid. Once installed, a PV module is very reliable because it has no moving parts and can operate for decades with no service required. In addition, a battery backup to the solar cells is convenient.

With the combination of the technology of PV cells and the earlier-discussed concentrated solar power, a worldwide market has developed for this type of renewable energy resource.

Assuming that technology continues to move forward at its current pace, and that a mere 100,000 homes install rooftop PV systems within the next ten years, it is reasonable to estimate that such systems will represent a total PV energy package in the United States adding up to more than one quad per year.

(d) Conversion Efficiencies

In terms of *quads*, we have shown (in earlier chapters) that the solar energy received by the earth is more than 5 million quads per year, whereas the total annual energy used by the world's people is roughly 450 quads. In other words, the total energy used by humans over the course of a year is delivered to the earth from the sun in approximately one hour.

Even though the incident solar energy is massive, very little of it is used to provide power for human activity. For example, although the best of today's commercial PV cells, using single-crystal silicon, are a little more than 15 percent efficient, it is still a fact that because of current availability and complexity considerations, it is currently simpler and more practical to obtain electricity by going through the conversion from fossil fuel burning to electricity production. We therefore find that less than .02 percent of world electricity production comes from solarvoltaic technology. This may change as the elements of nanoscience are further developed and understood.

With advances in such science, instead of depending on electricity, we may turn to solar protons producing forms of chemical energy through photosynthesis, where sugars and starches are produced in plants, simultaneously growing leaves, roots, and stalks. These products of nature lead to the development of biomass, a small percentage of which can be converted to power production through technology we have already developed.

A common attraction of solar energy is its use as a producer of heat, without going through any process of combustion. The unconcentrated energy can be used to heat space and water in many residential applications. When concentrated, via troughs, dishes, or towers, it can produce very high temperatures such as those required for powering steam engines. A recent report by the US Department of Energy noted that there is much room for solar energy improvement, since photovoltaic conversion efficiencies are generally less than 10 percent, photosynthesis conversion efficiency is less than 1 percent, and solar thermal efficiency is less than 30 percent.

4. Indirect Solar Radiation

When most people read about or make reference to solar energy, they are thinking about some of the examples of direct radiation that have just been described. However, energy from the sun is manifested on the earth in many different ways, even if the sun's role is not immediately obvious. *Indirect* solar power uses several transformations to convert the solar radiation to the usable form that we desire. For example, biomass, discussed previously, goes through photosynthesis to produce chemical energy that ultimately produces heat, electricity, and various biofuels. Following are several examples of indirect systems, such as *hydropower* and *wind*.

(a) Hydropower

Most of the earth's surface consists of ocean waters, so that a massive amount of solar energy received by the earth is collected in these waters. This energy is absorbed as heat, and it causes some of the surface water to evaporate and rise into the atmosphere as water vapor. This leads to the formation of clouds that float through the skies, ultimately condensing and falling back toward the earth in the form of rain. Much of this rainwater is accumulated in high-altitude regions of the world. Then, under the influence of gravity, the water flows down rivers and over dams, exerting energy that then can be captured and used for many purposes. The process of capturing and using this energy is known as hydropower. Second only to Canada, the United States is one of the largest producers of hydropower in the world.

Before going further on the use of hydropower, it is of value for the reader to understand what a turbine is and how it works. A turbine is a very simple device used by humans in many ways. It is basically a wheel turned by the force of a moving fluid pushing on its blades, similar to a pinwheel when you blow on it. The whirling turbine then spins an electric generator that produces electric power. Some turbines might be fueled by steam or gas. However, here we are considering *water turbines*, which may be located at dams and waterfalls and which require nothing more than a constant flow of water to spin the wheel and the shaft, which in turn spins the electric generator, thus converting the original energy to electric power, which is then transmitted through cables to homes and businesses. Modern hydro turbines are extremely efficient, converting 90 percent of the available energy into electricity.

Hydropower is currently the primary renewable energy source used to generate electricity. It is cheap and clean. The power produced by a water turbine depends on the amount of flowing water and the height that the water drops before hitting the turbine wheel. The water and air are not polluted at all. The only negative effect of the use of hydropower is that a series of dams have to be constructed, which makes it more difficult for those fish who swim against the current to get to their spawning grounds. After providing the desired hydrogen energy, the water then flows back into the oceans, from where the cycle is repeated, without producing any pollution or waste. Originally, hydropower was used for irrigation, water wheels for sawmills, and barge transport, but today hydropower is used primarily for electric power generation.

Some countries depend almost exclusively on hydropower to produce their electricity. In the United States, almost 10 percent of our electricity comes from this source. Norway gets 99 percent of its electricity from hydropower, and New Zealand gets 75 percent. In the rest of the world, hydropower produces close to 20 percent of electricity. Although it is efficient and cost-effective, less than 5 percent of this country's existing dams are being used today to generate power. US hydroplants vary in size from many small ones that power only a few homes to giant dams, such as Hoover and Grand Coulee, that provide electricity to millions of people. Over one-half of the total US hydroelectric capacity for electricity generation is concentrated in three states (Washington, California, and Oregon). A doubling in hydropower activity is proposed in chapter 5 as we reexamine our national energy policy.

The combination of earth rotation and lunar gravitational attraction has led to the phenomenon of tidal action, and the tides often are identified as an energy source. However, they may also be considered as a means for increasing available hydropower. For example, it is possible to construct dams that capture high tide water and then provide hydroelectric power when the water recedes. The best locations for using this energy source are where the tides are high, such as the Bay of Fundy off Nova Scotia or the Yellow Sea off the coast of Korea.

A principal benefit of hydropower is that it does not produce any harmful emissions such as CO_2. In addition, it is economically beneficial, even after the water-restraining dams have been built. As noted, only about 3 percent of the dams in the United States are used for hydropower, primarily because of concerns about the effects of reservoirs on the environment. However, installing turbines in existing dams can be a significant and cost-effective power source. In addition, small-scale power systems can be used in small rivers and streams merely through use of water wheels, without a need for dams or water diversion.

Hydropower faces some immediate problems, including environmental issues, regulatory complexity, and energy economics. Notwithstanding these concerns, however, it is estimated that the United States will be using hydropower to fulfill one quad of energy within the next decade and ultimately two quads by the end of the century.

(b) Wind

Winds are the result of solar energy combined with atmospheric structure and earth rotation. This leads to various temperature gradients and pressure differences. As the air in high-pressure zones rushes toward the lower pressure, we see the phenomena of "weather," including turbulence, calms, storms, rain, snow, and wind.

Wind was first recognized as a practical source of energy around the twelfth century, when the windmill was first used. Early Dutch windmills were used to pump water and to drain lowlands. Today they are used in America on a limited basis, but we do have tens of thousands of wind-driven one-kilowatt generators, used by farmers to trickle-charge twelve-volt batteries that provide farm lighting and power. For windmills to be useful and economical for such farms and private homes, wind speeds need to average nine miles per hour year-round.

Recent studies have shown that the use of the wind on a massive basis could provide a significant portion of the country's energy. This would involve constructing large windmills, perhaps 200 feet in diameter, spaced a few miles apart in high-wind areas, such as the Great Plains, the New England shores, and the Texas Gulf Coast. Wind energy is non-depletable and nonpolluting, although the large towers may be considered aesthetically undesirable.

NASA has been developing lightweight and efficient blades for windmill use. These may be considered for large power systems or for individual home use. The wind appears to be a clean, significant, and technologically available source.

Wind farms are groups of windmills scattered over a large area. Currently, many of these farms are owned by private individuals who sell the produced electrical energy to utilities. Americans originally used small windmills to pump water, grind wheat, saw wood, and provide limited electricity. Today there are enough windmills in this country to provide electricity for a city the size of Chicago, but this is less than 1 percent of the nation's electricity consumption.

There are a lot more windmills to be built if the national energy contribution is to be significant. At the present time, close to 17,000 megawatts of wind power capacity have been installed in the United States, providing annual electricity needs for close to 4 million average American households. It is estimated by the US Department of Energy that 6 percent of the country's contiguous land area has strong and continuous winds, with the potential to supply more than the national electricity usage. In terms of total wind energy capacity, the United States is the world leader. Cumulative world capacity is approaching 100,000 megawatts. At continuous full-wind production, this would be equivalent to three quads.

One attraction of wind energy is that the land on which windmills are located can be used simultaneously for farming, ranching, and other commercial activities. Wind is a fast-growing energy source that is clean, abundant, and relatively low-cost to produce. The blades of the windmill are attached to a rotor, and they capture kinetic energy from the wind, spinning a shaft that is attached to a generator. This produces electricity. The electricity can then be stored for later use in windless periods. Some of the electricity may be delivered to an inverter, which converts the DC current to AC current and boosts the voltage to a value that is convenient for the property owner. The wind turbine is most effective when mounted on a pole or tower at least eighty feet above ground level and above nearby obstructions or tree lines.

One challenge is that wind is not constant, and alternate energy sources should be available. The usefulness of wind power is significantly decreased in cities and suburbs, where turbulence created by windmills can be disturbing to nearby residents. Also, concern has been expressed about the appearance of giant windmills, the noise that may be generated, and the killing of birds who fly into the blades. This latter effect is negligible when compared to the billion birds a year who are killed by house cats in the United States. If a program were to be undertaken to build and group together new turbines varying from 100 kW to several megawatts, as well as single small turbines for homes, it is not unreasonable to expect that this source of energy would produce four quads per year within the next twenty years.

C. Nuclear Energy Sources

For a while during recent decades, there was a frequent feeling that the solution to the "energy problem" was for the world to shift to nuclear energy as its fundamental source of power. Recently, because of pressures from environmental groups and the general concerns about the safety of nuclear plants and the handling of nuclear wastes, there has been a halt to the growth of such activity, and the entire concept of nuclear power is being reexamined at both governmental and lay levels. The two primary means of deriving nuclear power are classified as nuclear fission and nuclear fusion. This section describes some of the fundamentals of these concepts, using some of the basic definitions presented in Appendix 1.

The structure of a typical atom is illustrated in Figure 3C-1. As shown in this figure, an atom is basically made up of electrons (negatively charged), protons (positively charged), and neutrons (no charge). The protons and neutrons are packed closely together in a nucleus. The electrons rotate about the nucleus in various orbits, or shells. The mass of a proton is approximately equal to that of a neutron, and the mass of an electron is so much smaller as to be negligible.

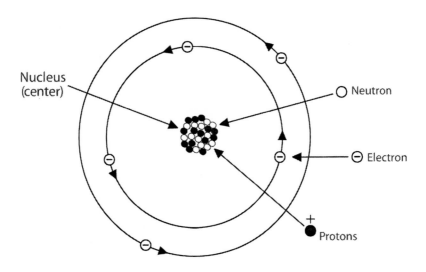

Figure 3C-1

Typical Structure of an Atom

Electron

Protons

These particles have positive electric charges (balancing out the electron charges). Each element has a different number of protons in it nucleus, and this defines the "atomic number" of the element.

These are extremely tiny particles (of negligible weight compared to the particles in the nucleus). Their distance from the nucleus is thousands of times greater than the diameter of the nucleus. Here they are shown rotating in two shells, but as the number of electrons increase, they will travel in up to seven shells.

Neutron

This is the other kind of particle found in the nucleus. It has no charge, but is about the same weight as the proton.

We can currently identify 116 known elements, most of which occur in nature, but a number of which have to be created chemically in order to exist. Elements are identified by their atomic number, which is equal to the number of protons in the nucleus (which in turn is equal to the number of orbiting electrons). Some of these elements, such as aluminum, fluorine, and phosphorus, can exist only with a specific given number of neutrons in the nucleus. On the other hand, most elements can exist with several forms of atoms called isotopes, where each isotope has a different number of neutrons. The name and number of the element does not change, but the mass of each of the isotopes is different. For example, hydrogen (the lightest element) is element 1, with one proton and one electron. However, hydrogen has three isotopes, with zero, one, or two neutrons in the nucleus. Its atomic mass may therefore be 1, 2, or 3.

Another element, which will be discussed in more depth shortly, is uranium (symbol U, number 92), one of the heaviest natural elements, with 92 protons in its nucleus and 92 orbiting electrons. When uranium is found in the ground in its natural state, it is made up primarily of two isotopes: one has 146 neutrons in its nucleus (atomic mass = 92 + 146 = 238), and one has 143 neutrons (atomic mass = 235). U-238 makes up more than 99 percent of natural uranium; U-235 is a rare isotope, representing less than 1 percent.

In chapter 1, we noted that the total quantity of universal mass plus energy is conserved. Energy and mass have a number of forms, and each can be converted from one form to another. In addition, Einstein showed that mass can be converted into energy. The concept of nuclear power is based on the recognition that if a tiny bit of mass can be made to disappear as mass and convert to energy, the amount of energy so created is tremendous. The idea can be realized by considering the equation that relates energy (E) to mass (m): $E = mc^2$. One can imagine how great E must be when it is realized that c represents the speed of light (186,000 miles per second). If the units of E, m, and c in this equation are all converted to familiar values, the formula relating mass to energy may be expressed in the following form: 1 gram of mass = 85.2 billion BTUs, or 1 quad of energy = 11.7 kilograms of mass.

1. Natural Fission

In prior years, the word "fission" was a biological term used to describe the splitting of one living cell into two. In 1940, the word was used to describe the phenomenon of the nucleus of the atom splitting into two smaller nuclei when it is struck by a fast-moving neutron, also releasing some of the neutrons that were contained in the original nucleus. This is illustrated in Figure 3C-2. Here we see three stages of a nuclear fission event. On the right, we see that the fission products include the two nuclei of two lighter elements, plus a number of free neutrons moving out in various directions. The total mass of all these products is a tiny bit less than the mass of the original heavy nucleus.

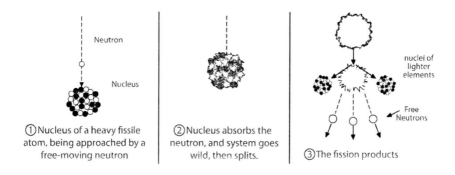

① Nucleus of a heavy fissile atom, being approached by a free-moving neutron

② Nucleus absorbs the neutron, and system goes wild, then splits.

③ The fission products

nuclei of lighter elements

Free Neutrons

Neutron

Nucleus

Figure 3C-2

A Nuclear Fission Event

Now let us consider that instead of a single heavy atom, we have concentrated this material into a dense mass, such that millions of these nuclei are close together, in what we call a critical, or supercritical, configuration. In this case, the three free neutrons in Figure 3C-2 will not all escape into space; one (or more) of these will collide with the nucleus of another atom, and the process will be repeated. This continues rapidly and is known as a chain reaction.

When there is a possibility of a chain reaction, the element is known as *fissile*. There are only a few chemical isotopes that are fissile, and these are known as nuclear fuels. Of these, there is only one found in nature that is capable of fissioning under relatively mild conditions, and that is U-235, an isotope of uranium. If a free neutron runs into a U-235 nucleus, the neutron is absorbed, and the nucleus becomes unstable and splits. This fuel is contained in a conventional fission reactor that is basically a large steel pressure vessel with a central core holding the U-235 nuclear fuel. If enough of this fuel is concentrated (such as a two-pound sphere of U-235), a chain reaction starts, with continuous splitting of nuclei, releasing of neutrons, and production of energy. Two pounds of uranium releases about 75 billion BTUs during the fission process. Therefore, the total energy obtained from the decay of a uranium sphere (via a chain reaction) is greater than the energy produced by millions of gallons of gasoline. This energy heats the surrounding water, which then creates steam and drives turbines to create electricity. After about a year's operation, the U-235 in the reactor is all converted or split into lighter atoms, and the fuel in the nuclear firebox must be replaced.

The concept of obtaining energy by splitting fissile atoms is attractive, but U-235 is the only natural source that can be used for this purpose. However, there are several other elements found in nature that have the capacity of being converted into fissile materials. To make this conversion, we must use a breeder reactor, as described in the next section.

2. Breeder Fission

U-238 is a much more common isotope of uranium found in nature, but it is not fissile. However, if bombarded by a host of fast neutrons, a change can occur within its nucleus, producing plutonium, element 94. One of the isotopes of plutonium is Pu-239, which *is* fissionable. In a specially designed breeder reactor, the isotope U-238 is combined with the U-235. In such a reactor, at the same time that energy is being obtained from the fissile U-235, other fast neutrons are released to change the U-238 to U-239, which ultimately becomes Pu-239. This is an isotope that is not found in nature, but is an ideal fissile element for energy production purposes.

Another similar element that has the capacity of being bred is thorium-232, which is also a natural element capable of being converted into a fissile material, in this case, the uranium isotope U-233. The potential availability of nuclear energy is therefore increased considerably if breeder reactors are used for the creation of these unnatural materials.

Based on the various estimates of the availability of uranium and thorium, regardless of the form or whether or not they are recoverable, it is reasonable to estimate that nuclear fission represents an energy source capacity of more than a million quads. This is equivalent to an unlimited supply for humans; unfortunately, however, there are a number of significant problems with the use of nuclear fission, all of which are associated with the risk of exposing people to dangerous radiation.

At the present time, there are more than 400 nuclear power plants around the world, primarily dedicated to producing electricity. About 100 of these are in the United States, and they provide about 15 percent of current US electricity.

If carefully constructed and operated, nuclear plants can be clean, and there is no carbon dioxide associated with this energy source. The plants can be sited anywhere, and there is no danger of running out of the supply. The concept of producing large quantities of energy by means of nuclear fission is attractive, but there is one negative aspect that is worrisome. After the fission occurs, the remains of the uranium nucleus are not split exactly in half. They may be any pair of 200 different isotopes, including cesium-137, carbon-14, radium, strontium-90, and others, as well as traces of the original fuel, which is plutonium or uranium in the breeder, and they may contain as much as 5 percent of the energy that is released during the fission process. That energy, now known as radioactive waste, appears in the form of radioactivity, of varying intensity and duration, and this can be neither controlled nor eliminated.

Therefore, the attractiveness of nuclear fission as a source of energy is tempered by the recognition that the waste products are highly radioactive and may continue to exist for periods of time varying from seconds to thousands of years.

The processes of mining, purifying, and transporting the nuclear fuel are all areas of concern, as is the proper functioning of the power plant. Therefore, even though nuclear fission is achievable and can satisfy humans' energy needs, the public concerns regarding safety and waste disposal have significantly reduced the probability of its general use. A large nuclear-waste storage facility has been built in Yucca Mountain, but there is no guarantee that this or any other nuclear waste storage facility will be completely safe for millennia to come. The casks that hold the nuclear waste must themselves withstand any possibilities of damage, and they must be shipped to Yucca Mountain via thousands of miles of routes that may endanger millions of people who live nearby.

Therefore, consideration must be given to the problem of handling waste material, including whether and where it can be safely stored away for long and indefinite periods. In addition, because Pu-239 is the primary element for the production of nuclear weapons, questions regarding terrorism, theft, nuclear blackmail, and the morality of nuclear war are all inextricably linked with the question of feasibility of nuclear fission. Such obstacles have resulted in a near-halt to the continued construction of nuclear-fission power plants.

A formal application has been submitted for the operation of the Yucca Mountain facility for nuclear waste storage, but it is not anticipated that the Nuclear Regulatory Commission will decide whether to approve or deny such an application for several years. On the other hand, France is conducting a successful nuclear program based on the reprocessing of used nuclear fuel. This can be a dangerous policy, and it is not recommended as a future US activity.

From an emissions standpoint, nuclear power is much cleaner than fossil fuel sources. However, it is also much costlier, involves extensive safety and security issues, and includes a radioactive waste problem that can be frightening. The energy content of known and available fissile material is in the order of thousands of quads. However, because of the possible risk associated with the waste disposal, there is a growing natural tendency to avoid pursuing this source aggressively and to seek alternate solutions.

3. Fusion

In the case of fission, we have seen that a heavy atom is separated into two smaller ones, and the total mass of the fission products is less than the original, so that a portion of the mass is converted to energy. On the other hand, in the case of nuclear fusion, two light atoms combine (or fuse) into a single atom. This time the new atom has less mass than the two original ones, so that some mass is again converted into energy. Nuclear fusion is the most promising, but simultaneously the most challenging, of all possible alternate energy sources.

As described previously, the various isotopes of different elements are identified by their numbers. In the case of hydrogen, however, the isotopes also have names: protium, deuterium, and tritium. Remember that hydrogen is the lightest of all elements, with a single proton making up its nucleus. There are no neutrons in the protium isotope. The deuterium isotope has a simple neutron attached to the proton in its nucleus. Tritium, with two neutrons, does not exist in nature, but it can be made.

If the nucleus of a deuterium atom can be combined with a tritium nucleus, the combination produces a helium atom with one extra neutron. This extra neutron is blown away with a tremendous amount of energy, and all that remains is a single and safe helium atom. There are no other new elements created, and no radioactivity or waste material to be a matter of concern. Fusion is therefore an attractive energy source, involving the combination of isotopes of some of the lightest elements (hydrogen, helium, lithium).

And now comes the problem: it is not easy to combine the nuclei of two light elements! Each has a positive electric charge, and therefore, as they approach each other, a tremendous repulsive force develops between them, so that there is little possibility that they will fuse. However, we know that fusion takes place in the sun, so that all that is needed is a dense contained gas, closely packed particles, and a very high temperature. One way to simulate this environment is with a powerful magnetic field. Obviously, the creation of such an environment will require a lot of energy, but it will pay off if the energy that is produced is far greater than that which is used in its production. However, obtaining a sustained fusion requires very large-volume high-density plasmas that reach temperatures of several hundred million degrees Fahrenheit and that hold these temperatures for several seconds. This is not easy and is beyond our scientific capabilities at the present time.

In each of the energy sources discussed so far in this chapter, there has been a description of the concept and an estimate of the amount of energy the source may be able to provide. Whether or not the cost is reasonable or prohibitive is a separate issue to be discussed later. The prime considerations in the concept of using nuclear fusion are as follows:

- The sources of fuel (hydrogen isotopes) are abundant.

- The primary waste material is helium, which is harmless.

- The fusion concept is understandable, and research into controlled fusion for the production of electricity has been going on for fifty years. Several fusion reactors currently are in experimental stages around the world, with a number of scientists who are optimistic that the technology will be solved in the future.

- Although the concept is understandable, science has not yet developed a way to achieve controlled nuclear fusion. The primary problem is that the temperatures and pressures necessary to achieve the fusion process are millions of times greater than humankind has ever used before. It may be that the goal is not achievable.

In view of this, we make no forecast at this time about the number of quads that we hope to attain through the use of nuclear fusion.

D. Geothermal

More than 4 billion years ago, a collection of tiny particles and gases combined to form the earth, a very hot sphere roughly 8,000 miles in diameter. Over time, the surface cooled, and an atmosphere formed; today the surface is made up of dry land and oceans, but the temperature at the center of the earth is still many thousands of degrees Fahrenheit. The outer layer, or crust, is three to thirty-five miles thick, with a number of places where cracks and faults periodically appear, allowing volcanoes, hot springs, and geysers to break through. Beneath the crust is a region of extremely dense rock material known as the mantle, where the rock temperatures are close to melting, between 1,000 and 2,500 degrees Fahrenheit. Below this we encounter magma, a mantle rock so hot that it is in liquid form, carrying up the heat from the earth's core, where the temperature exceeds 9,000 degrees Fahrenheit. By geothermal energy, we refer to heat and power that flows upward from the center of the earth and that we can use without concern about polluting the environment.

Magma rarely rises all the way to the surface, but when it does, it is generally in the form of lava, emitted from volcanoes. More commonly, it heats water that has seeped into the earth (from rain, oceans, etc.), much of which returns to the earth's surface as hot springs, geysers, and geothermal reservoirs. Such reservoirs may produce high-pressure steam that can spin electric generators or merely hot water that is useful, reliable, and clean.

To increase productive use of geothermal energy, we should identify those regions of the world where magma rises closer to the surface and be prepared to reach the geothermal reservoirs by drilling. There are many "hot spots" on the earth's surface, particularly at the boundaries of the plates that make up the earth's crust and at places in the crust that fracture periodically. At present, only a small portion of them are being used as geothermal energy sources. This can be increased enormously with the application of new drilling technology, going down two miles into the earth.

In the year 2000, 9,000 megawatts (MW) of electricity were produced by 250 geothermal power plants operating in twenty-twp countries around the world; of this quantity, 3,000 MW were produced in the United States. An additional 2,000 MW in the United States represented non-electric geothermal energy used for various heating systems (for residences, spas, industrial plants, and greenhouses). At present, this is far less than one quad per year, but with concentrated development, geothermal production can be very significant in the next decade. In addition, the energy is clean, reliable, and flexible and requires less land than any other energy sources: no dams, forests, tunnels, mine shafts, or open pits.

When society chooses to make a significant effort to utilize geothermal energy, it will be divided into the following three categories:

1. Shallow Hydrothermal Systems

Shallow hydrothermal systems are employed when the magma has permeated the upper crust but not yet broken through. The hot rock is permeable, so the water or steam circulating through it is extremely hot and can be withdrawn, with temperature ranges between 100 and 300 degrees Fahrenheit. As the water or steam is removed, new heat flows into the reservoir from the depths below. This energy is used primarily for the generation of electricity. The steam and/or hot water used for this purpose is the prime element of geothermal power plants.

A somewhat more attractive resource exists if injection wells are drilled farther down to two or three miles. Injected water temperature is brought to 300 degrees Fahrenheit, and the water is then returned to the surface through production wells. Hydrothermal systems are currently in an early development stage.

The advantages of geothermal power plants are that they take up a very small amount of land area, are very reliable, run twenty-four hours per day, can easily be increased in size, and are not dependent on any fuel importation. At the present time, the United States is producing about one-third of a quad per year of geothermal-based electricity.

2. Direct Use

In many areas of the globe, the geothermal water may not be hot enough to be ideal for electricity production, but it can be an ideal hot water source for home heating and industrial processes. In such cases, a well can be drilled to extract the hot water for direct use in a mechanical heat exchange system. In addition to the heating of buildings, there are many opportunities for such applications, including crop treatment, horticulture, hot springs and health spas, fish farms, greenhouses, and snow and ice treatment, but intensive programs have not yet been generally undertaken for these applications. On the other hand, the city of Reykjavik in Iceland uses geothermal energy as its prime source of heat, and it has now become one of the cleanest cities in the world, even though it is also subject to volcanic eruptions. In earlier times, the air in Reykjavik was highly polluted by fossil fuel emissions.

At present, there are hundreds of communities in the United States that use geothermal resources. The one-third of a quad of electrical energy referred to previously is based on the use of known "shallow" hydrothermal reservoirs. If, in addition to these, we undertook a national program to drill deep holes into hot dry rocks and use these as heat sources, several quads of geothermal energy could be captured and used.

3. Geothermal Heat Pumps

Geothermal heat pumps, also known as ground-source heat pumps, use the natural temperature of the earth instead of producing heat by burning fuels. Since the interior of the earth is at a constant temperature throughout the year, the heat pump can exchange stored heat to provide warmth in the winter and transfer heat to the ground in the summer, and it can work well in crop areas, forests, and deserts. In addition to heating and cooling the air as required, it can also be designed to provide the home with hot water. The initial cost of such a geothermal energy system is high, but the costs are generally returned to the homeowner in less than ten years, and the system life is estimated by the Geothermal Energy Association to be about twenty-five years. There is no requirement for any electricity to provide heat; electricity is needed only to move the heat by means of the pumps. There are approximately 50,000 geothermal heat pumps installed in the United States each year, and the potential for developing geothermal energy systems at an acceptable cost can ultimately be a large number of quads per year.

E. A Summary of Currently Available Energy Quantities

In reviewing this chapter, we must bear in mind that the current energy demand in the United States is in the order of 100 quads per year (from chapter 2, section C), but national as well as worldwide energy use is probably going to continue to grow. Table 3E-1 presents a brief summary of the energy sources that currently provide this nation's 100 quads, with comments showing why these numbers will change drastically in the next century.

Because of the anticipated development of hydrogen energy technology, fuel cell applications, hybrid cars, and so on, there will be continuously increasing efficiency in energy use, thus reducing some of the effects of population and economy growth. As shown in this table, most of today's energy supply comes primarily from fossil fuels; these will start running out in the next few decades and will be gone in a few hundred years. By that time, the nation's plans should be to obtain our annual quads from a different distribution, with concentrated thermal energy from the sun being our primary national source. If other countries in the world adopt a similar philosophy, world society can continue to prosper without depending on fossil fuels such as oil.

Table 3E-1

Energy sources	US current consumption (quads/yr)	Comments
Fossil fuels (coal, oil, gas, shale, tar sands)	89	We project to run out of coal in 230 years, and the others in less than 100.
Other solar sources (including biomass, hydropower, wind)	5	The supply and availability are far greater than current or future needs.
Nuclear	5	Fission produces dangerous waste; fusion may not be achievable.
Geothermal	<1	The potential available must be defined.
Total	100	Requires a sensible energy policy.

Chapter 4—Interim and Alternative Energy Sources

A. Electricity

Electricity is mentioned briefly in chapter 1 as a form of energy. But it is not one of the primary natural sources discussed in this chapter. Instead, it is an extremely convenient way of collecting energy from one of the primary sources and storing it for an intermediate period until we are ready to adapt it to a desirable application. Electricity can provide society with light, heat, and power. It is used to run home appliances such as washing machines, vacuum cleaners, ovens, refrigerators, and air conditioners; in industrial applications such as lathes, cranes, escalators, and computers; for transportation devices such as automobiles, trains, and boats; and in communication systems such as radios, TVs, and telephones.

It is much easier to use electricity as an energy carrier from power plants to homes and businesses than to use the original energy sources such as coal and oil. The current sources of US electricity are approximately as follows: coal, 52 percent; nuclear, 20 percent; gas, 16 percent; oil, 3 percent; hydropower, 7 percent. Annual electricity generation in the United States is presently in the order of fourteen quads.

One way to store electricity is through the use of batteries. These are devices that contain two different electrodes, such as plates. The electrodes have been processed so that one of them has an excess of electrons, which makes it "negatively charged," whereas the other has a deficiency in electrons and is therefore "positively charged." The difference in atomic makeup between the two electrodes represents chemical energy. When a wire is connected between the two electrodes, current flows through the wire, and the chemical energy is converted to electrical energy.

The voltaic cells discussed earlier, in chapter 3, section B-3(c), are an example of the direct conversion of solar energy to electricity. However, the most common way to produce electricity from a number of sources is by means of an electric generator. This involves whirling a magnet past coils of wires or copper bars. To produce the kinetic energy to cause the whirling, one might use the burning of gases, the concentration of heat, or hydropower from dams.

This technique for generating an electric current was developed in 1831 by Michael Faraday, an English scientist. His work involved spinning a small magnet inside a coil of copper wire, producing tiny electrical currents and creating an electrical charge. Fifty years later, Thomas Edison enlarged this activity tremendously by combining a large direct-current generator with a steam engine, thus producing large-scale electricity generation.

Electricity generation is the process of converting some of the natural sources of energy into electricity. This is done in this country primarily by burning fossil fuels, especially coal. The coal is dug out of the mines and transported to a steam-powered electricity-generating plant. There the fuel is burned to produce steam, which then rotates turbines to produce electricity. This electricity can then be transmitted through wires to serve the needs of businesses, industries, and homes in the area serviced by this particular generating plant. Such generating plants, distributed throughout the country, currently provide more than 70 percent of the electricity used in the United States. In addition, most electricity-generating facilities are large central stations occupying many acres of land for the power plant operation, the fuel storage facilities, and the structures for connecting to the transmission grid. Much of this land is scarred, polluted, and permanently unsuitable for much future use, following its normal life of forty to sixty years. Additional long-lasting land impacts include toxic residues, solid waste storage problems, and underground contamination possibilities.

The use of electricity for providing heat is inefficient and sometimes irrational. Heat is initially available through direct solar radiation or the storage of biomass energy such as wood, which then can be burned to provide warmth. It is not very logical to burn a natural energy source in order to convert it to electricity that then must be converted back to produce heat.

A significant portion, but less than 10 percent, of US electricity is produced by nuclear fission. Here, there is not any pollution from the production of greenhouse gases. However, as discussed earlier, the waste products of nuclear energy are themselves radiating and dangerous, and it would not be wise to continue to generate electricity in this way.

This reduces our sensible choices of original energy sources to those that are renewable and nonpolluting. Such energy sources include geothermal, hydropower, wind, and solar. Based on the application of near-future technology forecasts, availability, economics, and common sense, it is shown in the chapter 5 that the use of solar energy is a predominant choice over all the others, and the technology for using it is described there.

B. Fuel Cells

Early in the nineteenth century, it was known that photosynthesis is a process by which water (H2O) could be split into hydrogen and oxygen by using the energy from sunlight to combine with the chlorophyll in green plants. In 1839, Sir William Grove developed a concept for reversing the procedure, known as a *fuel cell*. In the 1960s, fuel cells were introduced into the US space program as power producers, and today they provide electricity and water to the space shuttle.

Like electricity, fuel cells are not original sources of natural energy, but instead they combine original energy forms to generate power in ways that are convenient and efficient. Basically, a fuel cell is an electrochemical device that combines hydrogen and oxygen to produce electricity, heat, and water, with zero or a minimum of polluting byproducts. The primary element of fuel cells is hydrogen, which may be supplied as a pure element, or as a part of methanol/ethanol, gasoline, diesel fuel, or ammonia.

Figure 4B-1 shows an elementary form of a basic fuel cell (commonly known as a proton exchange membrane fuel cell). The fuel, fed into the anode, may be pure hydrogen or any hydrocarbon (which contains hydrogen) such as gasoline, methanol, or natural gas. Oxygen (or air, containing oxygen) enters the fuel cell through the cathode. With the help of a catalyst in the anode, the hydrogen atoms are "split." The electrons travel through a separate path, and the electrical energy then can be used to satisfy the user's needs. This is the primary direct current that can be used to power motors or any other electrical devices. At the cathode, the electrons and protons recombine with the oxygen, and the primary product is water, with no combustion and little air pollution (zero if the fuel is solely pure hydrogen).

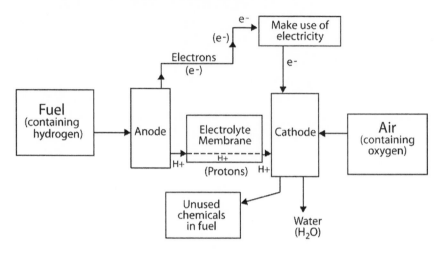

Figure 4B-1

Basic Fuel Cell
Also known as a Proton Exchange Membrane Fuel Cell (PEMFC)

Fuel cells are currently in use for vending machines, highway road signs, and many buildings such as hospitals, schools, and utility power plants, where they can provide backup for critical areas. They make no noise, and they are lighter and far more efficient than batteries. The attraction of fuel cells is that they contain no moving parts, produce no pollution, are extremely efficient, and do not depend on fossil fuels. In addition to the stationary operations, fuel cells are extremely reliable in transportation devices, such as planes, trains, and boats, and for telecommunications, portable power, and consumer electronics. At the present time, all the major automobile manufacturers are involved in development programs to commercialize autos using fuel cells as their power source.

In the discussion of hydrogen that follows, it is noted that hydrogen does not exist alone in nature as a pure element, but rather is always in combination with other elements, unless steps are taken to separate it. In applying pure hydrogen to a fuel cell application, there will be zero pollution. The only byproducts are electrical energy, heat, and water. If, on the other hand, fossil fuels are used as the hydrogen source fed to the anode, there will be carbon emissions, but even this would produce less than half of the emissions of combustion.

At present, a significant amount of research is being done on the optimization of fuel cell usage. This includes the development of low-cost electrolyte membranes; maintenance of a constant ratio between the electrons, protons, and oxygen; temperature management throughout the cell, and extension of durability and service life. In the case of automotive fuel cells, a minimum life span of 5,000 hours (equivalent to 150,000 miles) is desirable.

Fuel cells essentially have no moving parts and no combustion, and therefore they can be applied with high reliability to spacecraft, remote locations, power plants, and automotive vehicles. Further reference to this is made in section D-4 of this chapter.

C. Hydrogen

Just like electricity, pure hydrogen must be produced by using real initial energy sources, but then it becomes an extremely valuable interim source for enabling society to satisfy many of its energy needs.

Hydrogen is the simplest and lightest of all elements known to man, and it makes up more than three-quarters of the mass of the entire universe. It is colorless, odorless, tasteless, and nontoxic. It has an atomic number of 1 (one proton in its nucleus, and one electron orbiting around it). On earth in its natural state, it does not exist alone but is always combined with other elements. Combined with oxygen, it forms water; with carbon, it is found in all forms of life, such as plants or animals. It also combines easily and often with nitrogen. Here, we first consider how pure hydrogen is made, and then we examine how it can be used to our energy advantage.

The most common way to separate hydrogen from other elements is by steam reforming, in which the hydrogen atoms are separated from the carbon atoms in natural gas, such as methane (CH_4); unfortunately, the carbon is then free to join the gases that pollute the atmosphere.

A much more environment-friendly (but much more expensive) process is electrolysis, which splits water into hydrogen and oxygen.

As its name implies, electrolysis involves the use of electricity, and this offers no advantage if the electricity is generated using fossil fuels, since this generates pollution. Solar energy is the prime source for electrolysis, thus producing no unhealthy emissions, and the technology is being intensively researched to minimize the costs of the process. After it is separated, the hydrogen can be stored or transported to wherever and whenever the energy is needed.

Hydrogen has the highest energy content of any common fuel by weight, and it can be stored for long periods of time. When it is burned (combined with oxygen), it releases much energy, and the end product is water, with no concerns about carbon or sulfur dioxide. Millions of tons of hydrogen are produced annually in the United States at the present time, and this amount can be increased significantly. Hydrogen is not only an attractive storage medium, but it is also convenient for the transmission of energy over long distances. Just as high-voltage electricity can be transmitted through wires, hydrogen can carry long distances through pipelines. Such pipelines must be designed to avoid leaks and use materials that are compatible with hydrogen transmission.

One major application would be the development of fuel cells for principal use in powering vehicles. As described herein, the fuel cell is a device that can produce electricity by chemically combining hydrogen and oxygen, using the electricity to power a motor. The hydrogen is stored in the vehicle, and the oxygen comes from the air.

A disadvantage is that the storage of hydrogen requires a very large tank, even if it is stored as a compressed gas or a cryogenic liquid. Hydrogen is fed into the "anode" (one of the electrodes) of a fuel cell, where its atoms are split into electrons and protons. The electrons are then blocked and channeled through a circuit to produce electricity. Oxygen from the air enters the cathode (another oppositely charged electrode) and combines with the electrons and protons to form water.

At present, a number of organizations throughout the world are intensively involved in the production of hydrogen through electrolysis, using sunlight as the only energy source. With a solar-to-hydrogen conversion efficiency of 15 percent, the solar radiation over an area of 70,000 square miles (less than 1 percent of US land area) could supply 60 percent of the energy needs of this country. The global warming threat could disappear, as well as respiratory diseases produced by polluted air.

Because of the massive availability of solar energy and the ability to convert it into hydrogen as a principal carrier, many experts project that our current "fossil fuel economy" will eventually be replaced by a "hydrogen economy." One of the attractive advantages of the hydrogen economy is that hydrogen can be produced with simple technology anywhere that we have electricity and water. In the plan described in the next chapter of this book, a transition to a hydrogen economy is projected to be effective before the end of the twenty-first century.

D. Automobiles without Oil

Among the points earlier made in chapter 2 are the following:

- The measurable world energy use is about 450 quads. Of this, the amount used in the United States is about 100 quads.

- The amount of US energy consumption that depends on oil is 39 percent.

- In the "transportation" category, two-thirds of the energy is used by automobiles and trucks, with the rest devoted to airplanes, trains, and ships.

Considering these statistics in more detail, one can conclude that more than thirty quads of US energy expenditure is devoted to the use of oil as a fuel for cars and trucks. Therefore, let us examine the common automobile and determine what actions can be taken to satisfy all our transportation needs without the use of oil.

1. What Makes Conventional Automobiles Run?

When an automobile is driving along a road, it is exerting kinetic energy, and the source of this energy must come from a substance that we put into the car and that contains a great deal of chemical energy. The most common substance used for this purpose is gasoline, a hydrocarbon made from petroleum, often referred to as "oil."

There are a number of other possible fuels that can be used and that are not made from petroleum. These include alcohols such as ethanol and methanol, hydrogen, biodiesels, natural gas, and liquids made from coal. Some of these options are discussed in the pages that follow.

The hydrocarbons that comprise petroleum contain different numbers of carbon atoms in their molecules. The lightest gas in petroleum is methane (CH_4), which contains only one carbon atom. As the number of carbon atoms increases, so does the boiling point, and the molecules can be separated in an oil refinery through the process of distillation. The lighter molecules are mainly gases but also can be used in dry-cleaning fluids or paint thinners; the heavy hydrocarbons can be used as heating fuels, lubricants, greases, petroleum jelly (e.g., Vaseline), and paraffin. The intermediate hydrocarbons (like heptane, octane, nonane, and decane) are the ones that can be used to produce gasoline. These fluids can then be blended to produce regular, mid-grade, or premium gasoline, depending on that which is suited to perform best in the car engine design.

The ideal automobile engine is one that would burn the gasoline perfectly, providing the required energy and producing nothing in the exhaust except water and carbon dioxide. Unfortunately, such perfection does not exist, so the normal exhaust also contains polluting elements, such as carbon monoxide, nitrogen oxide, and some unburned hydrocarbons. These produce smog and greenhouse gases and may be the sources of unpleasant climate changes over the world.

The automobile was introduced to the world less than 200 years ago, in the form of a steam-powered carriage. Then, near the end of the nineteenth century, it was realized that the car energy could be provided through the use of electricity, in the form of batteries. Such cars, however, could travel only about fifty miles before their batteries had to be recharged, and these electric cars gradually lost favor. The gasoline-powered car was then developed, and by the beginning of the twentieth century, assembly-line production was undertaken. Within the next hundred years, automobile manufacturing became one of the largest industries in the world, and today the use of gasoline-powered automobiles is practically a modern necessity. But now, with the recognition that our current automobile fuel is going to be increasingly limited over the next century, it is time to consider other ways to make our automobiles run.

2. The Next Step: Hybrid Cars

As discussed previously, a primary usage of world energy is for transportation, in which the energy is converted from an original source (e.g., gasoline) into kinetic energy, which provides motion to vehicles such as automobiles, airplanes, buses, and motorcycles. This does not apply to bicycles, skates, or scooters, in which cases the source of kinetic energy is the human body of the rider. Transportation using conventional gas-powered vehicles accounts for 40 percent of all fuel use in America. In addition, about one-third of the country's air pollution comes from the tailpipes of today's vehicles. Instead of using conventional power sources based on gasoline engines, one can give consideration to using battery power, hydrogen fuel cells, or hybrid vehicles.

There are a number of current applications in which vehicles are powered solely by electric batteries, operating on the basis of stored electrical energy. Such batteries are many times heavier and use far more space than gasoline tanks and are practical for use only when the vehicles are used to travel short distances and can be recharged regularly (such as the little old lady who drives her car to the shopping center once a day for a ten-mile round trip, recharging the batteries at night). A more promising use of batteries applies to the hybrid vehicle, currently being introduced into the marketplace.

In the case of the conventional automobile, we are aware that its kinetic energy is converted through an internal combustion machine (engine), from the chemical energy in the fuel to the motion of a vehicle and its passengers. However, no one gives much consideration to where the energy has gone after the destination is reached and the automobile is parked. We do know, from the law of conservation of energy, that it has gone somewhere, possibly in some new form. The answer is that it has been converted primarily to the form of heat, warming the air through which the car has moved, the road along which it has been driven, and the tires and brake mechanisms that have been subjected to lots of friction. The energy is there, but it has been dispersed.

A concept for recovering and reusing some of this kinetic energy was developed in the last century, with the idea of mating the gasoline engine to an electric motor, which was essentially a generator that would recharge the car batteries. These batteries, when not needed elsewhere, could then provide electric power to help run the car as a supplement to the supply of gasoline. This was the beginning of the hybrid vehicle.

An early dictionary definition of "hybrid" was "the offspring produced by crossing two individuals of unlike genetic constitution." This has been expanded to refer to animals, plants, and more recently, hybrid automobiles. There are a number of ways of combining two different propulsion systems in a single vehicle; the most common is the gasoline-electric hybrid, which is powered by both a conventional internal combustion engine (ICE) and an electric motor. The ICE feeds on liquid fuels such as gasoline (or non-oil-based substitutes, such as biofuels), and the motor is fed by electric batteries. In the various hybrid designs that are presently available, the electric motor alone can provide low-speed operation without the use of the ICE and can provide supporting power when needed for accelerating or climbing hills. During conventional braking or normal coasting, the energy from the wheels is used to recharge the batteries, so that there is usually no need for a plug-in recharge. Another attractive feature of the hybrid is that the batteries take over and shut off the engine at stoplights and during stop-and-go traffic. This can save a lot of wasted energy.

Today, a hybrid vehicle generally refers to gasoline-electric hybrid vehicles, which use gasoline or diesel to power electric motors. Any time the car is braked, or is cruising or idling without the gas pedal being pressed, the batteries charge. The combination of electric motor and batteries provides sufficient energy to enable the vehicle to increase its mileage substantially for each gallon of gas that is used.

Among the principal features offered by hybrid cars are the following:

- When the car stops, the gasoline engine automatically shuts off. It then starts up again immediately, as soon as the driver steps on the gas pedal.

- When the brake pedal is stepped on, much of the energy normally lost in heating the brakes goes into the electric motor to help stop the car, and later to help drive it. This is known as regenerative braking.

- The electric motor and battery pack provide substantial help to the engine to accelerate the vehicle.

- At some speeds and driving conditions, such as when the vehicle is coasting or moving downhill, the vehicle depends solely on the electric motor and batteries, without using the combustion engine at all. This is known as silent cruising.

- One optional feature, now being offered in a number of hybrid vehicles, is the ability to recharge the vehicle from a home or garage electricity outlet, providing sufficient electrical energy to drive the car for up to sixty miles without using the internal combustion engine at all.

It should be noted that there are no claims that hybrid vehicles eliminate or dramatically reduce the number of quads of transportation energy used by society. What they *do* accomplish is a change in the *source* of the energy. The internal combustion engines in hybrids are smaller and use less fuel, and thus they produce less gaseous emissions into the atmosphere. The batteries now provide a significant portion of the vehicle's energy needs, and they may require periodic charging, but this is now *electrical* rather than petroleum-based energy. Of course, we must still consider the source of our electrical energy; this is covered later in this chapter.

As we move from today's conventional vehicles to those of the future (hybrids, pure electrical, fuel cells, hydrogen, etc.), we will be making continuous improvements in the development of a non-oil society. Instead of viewing gasoline and diesel as their sole fuel sources, automobile manufacturers are now giving consideration to biofuels, on a limited basis. Chapter 7 presents a time line with significant goals in the establishment of an energy-efficient national (and ultimately world) community.

The first hybrid vehicles were introduced in the 1990s, and now there are roughly 1 million hybrids on American roads. This is still less than one-half of 1 percent of all automobiles in America. However, it is a positive step being taken to increase fuel economy, and we can expect that marketing strategies and the potential for savings at the pump will make the adoption of hybrids an initial stage in a national program to establish energy efficiency. Within the last decade, we have seen the development of the Honda Insight and the Toyota Prius, as well as plug-in hybrid electric vehicles, discussed in the next section.

At the present time, companies producing hybrid automobiles include Ford, Honda, General Motors, Mazda, Renault, Toyota, and Lexus. In addition, a number of trucks, trains, and buses are being considered for conversion to hybrid technology. At the same time, much new technology is under way to develop more efficient and more sophisticated versions of the first hybrid vehicles. These include the use of hydraulic pumps, compressed-air power sources, diesels, and steam engines.

3. Electric Cars

Although hybrid cars improve the efficiency of mileage per gallon of fuel oil, consideration should be given to the use of vehicles that do not have any internal combustion engine, but that are fully powered by electricity alone. Such vehicles, sometimes known as BEVs (battery electric vehicles), are being used for golf carts, for forklifts, and on a limited basis for automobiles and light trucks.

Approximately a dozen companies are now developing and introducing electric cars, and we soon will see a number of them. Some of these can reach top speeds of 80 miles per hour, and the average range is generally less than 75 miles, but the latter will probably soon grow to over 100 miles on a fully charged battery. Today there are more than 50,000 electric cars operating in the United States, and Nissan's president is predicting that close to 10 percent of cars sold by the year 2020 will be electric vehicles.

A key attraction of BEVs is that they do not emit any exhaust products, but they do call for a general increase in electricity generation. One of their negative characteristics is that they are limited by the life of their batteries, which are one of their most expensive components. The batteries need to be replaced or recharged frequently, and recharging techniques are currently being investigated. One attractive feature of the batteries is that they can be partially recharged during braking of the automobiles, when they can recover much of the energy that would otherwise be lost through friction and heat generation. However, they still will require periodic recharging if the vehicles are to travel very far. In many cases, the recharging can be done while the car is in the garage at home, particularly when the travel is local. In other instances, battery replacement and/or recharging can be done at conventional gasoline stations, but this may take time (e.g., from ten minutes up to an hour), and it requires that battery service facilities be made available at normal fueling stations throughout the country.

As the use of batteries increases for electric car application, it is expected that advanced battery development will take place. Currently China is building dozens of advanced battery plants, whereas the United States has not given it much attention. We should expect to see increasing effort devoted to the development of light, cheap, long-lasting batteries for automobile application. According to recent statistics, American cars are driven about thirty-five miles per day on average. For families owning two cars, it would be attractive for one of them to be an electric car, used for short trips and recharged overnight.

An interesting and logical question relates to the source of the new electrical energy that is required for these cars. For example, if this energy were to be produced from fossil fuels, nothing would be gained by going to a conversion through electricity because we would still be using up the oil and creating the pollution. We will soon see that there are other ways to develop the electricity, such as where the original energy sources are solar rather than oil-based.

4. Fuel Cells in Automobiles

The electric cars just described obtain their energy from the common battery, which stores energy in chemical form. When the battery is called on to provide energy, these chemicals produce electricity, until the stored chemicals have all been converted to gas, and the battery must then be replaced or recharged. Even when it is operating efficiently, a car battery as the primary automotive source is much larger and much heavier than a gasoline tank.

A different electrochemical device is the fuel cell, described in section B of this chapter. This also converts chemical energy into electricity, but here the chemicals flow continuously into the cell, and there is no requirement for recharging. Fuel cell designs being considered are those that use hydrogen or a hydrocarbon as their primary chemical. These are processed to combine with oxygen (from the air), creating an electric current that powers the vehicle's motor. Hydrogen fuel cells produce no pollution, and their only byproduct is water. If the fuel is hydrocarbon, the waste products include carbon dioxide as well as water, but the level of pollution is significantly decreased compared to current values.

The first vehicles powered by fuel cells have recently been built in Denmark and can be used to haul luggage and provide short-range transportation in such places as airports, hospitals, and golf courses. The hydrogen is stored in pressurized canisters that provide a supply permitting fifteen hours of continuous driving, after which the canister needs to be replaced.

The major technological problem to be resolved in developing fuel-cell vehicles is finding the optimum way of isolating and storing hydrogen. As noted earlier, steam reforming and electrolysis are two of the techniques for obtaining elemental hydrogen and storing it in large quantities for application to vehicle fuel cells. A number of these techniques show significant promise but still require much development and testing. This will take time and effort but will be a major factor in the solution to the world energy problem. Chapter 5 discusses the transition of excess solar energy to create non-polluting hydrogen.

5. Pure Hydrogen Vehicles

An ultimate goal in making the gradual transition away from gasoline for transportation purposes is to ultimately obtain our vehicle power from pure hydrogen.

The concept of developing hydrogen vehicles presents several challenges; one is that of producing pure hydrogen, and one is the problem of storing it. Before starting to produce hydrogen, we must recognize that it is *not* a source of energy. It is a *carrier* of energy, similar in some respects to electricity. One way to accumulate hydrogen is to use a process called reforming, which separates it from the hydrocarbon such as methane. In the process of reforming, however, a lot of carbon dioxide is created. Reforming also requires the use of very high temperatures and expensive catalysts. A better procedure is to split water using electricity (electrolysis), producing only hydrogen and oxygen atoms. The hydrogen is then carrying a great deal of energy, and when this is combined in the fuel cell with oxygen, the recovered energy is available to drive the motor and hence the vehicle.

This leads to an interesting question: if we need a source to transfer energy to the hydrogen and later convert the energy from the hydrogen (in the fuel cell) to drive the vehicle, why not use the original energy source and skip the hydrogen entirely? The answer is based on the following facts:

- The energy source easiest to use is gasoline, but it will all be gone within this century.

- While being used up, gasoline is dirty and polluting.

- We need an energy source that is equal to or greater than our continuing transportation needs and that is preferably clean and easy to use (once it is established).

- The only obvious and available source that meets the foregoing requirements is solar. The chapters that follow present some of the elements that make the solar-hydrogen economy a sensible basis for future transportation.

Before going on to the creation and use of pure hydrogen in a vehicle, let us consider the question of storing it:

One of the significant areas of current research is hydrogen storage. Hydrogen is a very light gas, and a significant amount of it is required to provide a vehicle with a long-range capability. Compared to the volume of a gasoline tank, the equivalent energy in normal hydrogen would require a tank with a volume thousands of times greater. This can be resolved a great deal by compressing the hydrogen gas, but significant pressure means that the tanks have to be thick and heavy, sometimes doubling the weight of the normal vehicle.

An alternative approach is to limit the distance traveled on a full tank of hydrogen. Standard ways of storing hydrogen on board are as a compressed gas, as a cryogenically cooled liquid, or in a hydride compound that releases hydrogen when reacted with water. In any of these cases, a large tank is required, and solutions to this problem need further investigation.

In the interim, we can generally anticipate that the maximum distance we can travel with a full supply of hydrogen is not much more than 200 miles. Therefore, it will be necessary to replenish the supply in our garage or look forward to hydrogen pump outlets at each commercial gas station, where we can refill our cars more frequently. Much research is currently under way to increase the amount of hydrogen that can be stored on board using high-pressure or cryogenic liquid hydrogen. In the meantime, the hydrogen-fueled vehicle is practical where only short distances are required, such as golf carts, food and laundry wagons, or maintenance vehicles in parks.

Based on the foregoing logical arguments, it is projected here that pure hydrogen will be the principal vehicle fuel by the middle of the twenty-first century. In the meantime, we can gradually convert our automotive propulsion systems from oil to natural gas to hydrogen at the same time that we go from internal combustion engines to batteries to fuel cells.

At the present time, GM has developed a hydrogen sport-utility vehicle presently called the Sequel, which has a range close to 300 miles and can accelerate from zero to sixty miles per hour in less than ten seconds.

It is expected that fuel-cell car engines will be less expensive and far more efficient than today's internal combustion engines. Returning to the phrase "a solar-hydrogen economy," we may conclude that transportation electricity in the next century will be provided through the use of pure hydrogen, as a carrier of energy received from the sun, and this energy will be used efficiently, safely, and cleanly.

Chapter 5—Secondary Elements of a Sensible US Energy Policy

A. Where Do We Go from Here?

So far in this book, we have discussed energy forms, energy sources, and the ways in which energy is needed and used in the recent past and the present. During the 1970s, the "energy crisis" became a leading subject of cocktail party. The news media bombarded us with energy analyses, philosophies, and dire predictions. These generally obscured our understanding of energy and mixed it up with discussions of oil reserves, nuclear policies, and regulatory practices. Recently, we have seen a shift from energy "crises" to oil "prices," which now holds a position as a high-priority item for political discussion. The "energy crisis" does not refer to a shortage of energy for satisfying our basic needs; it refers to our inability to satisfy our energy hunger at an economic price that we are willing to pay.

The world's population has been increasing dramatically and is expected to continue to grow. At the same time, we are undergoing technological developments that enable us to accomplish things that were never previously considered, and people now have the understanding and knowledge to expand our activities in communication, travel, style of living, recreation, scientific development, and satisfaction of all our needs and desires. In order to meet these objectives, the use of energy on an individual and a world basis is expected to grow significantly.

The expansion of energy usage can become a serious issue when consideration is given to the various energy sources that have been discussed in the previous pages. Other than solar energy, these sources may be considered as "secondary elements," since they do not have the

101

potential to grow at the rate needed by society over the next century. For example, the availability of fossil fuels will decline and disappear, and the costs will skyrocket as the fossil fuels are used up, while most recoverable energy sources are limited either in scope or in reasonable cost. Each of these "secondary elements" is discussed and analyzed in this chapter.

The primary and obvious energy source to satisfy all the world's needs is the sun. We will never have a problem of insufficient energy as long as the sun is present. The challenge of society is how to use the sun in an efficient and economical way. The use of solar energy is treated in chapter 6, and an overall plan using all these energy sources in a practical and economical manner is the subject of chapter 7.

B. Plan for Reducing the Use of Fossil Fuels

1. Oil

The distribution and use of specific sources of energy are discussed in various places in this book. In reference to fossil fuels, the three most common in the United States are oil, coal, and natural gas. Let us first give consideration to the largest of these: oil.

Most of the 100 quads of energy expended annually in the United States is devoted to transportation, so that our total annual use of oil for this purpose is approximately 40 quads. Based on the use of refined petroleum as our transportation source, and equating 1 quad to 172 million barrels of oil, the United States presently depends on 7 billion (40 × 172,000,000) barrels of oil per year, primarily to power our automobiles and trucks.

The use of oil has recently become quite unattractive, following a major accident that occurred during underwater drilling in the Gulf of Mexico in 2010. This has created a significant environmental concern that the collection of oil may represent a formerly unrecognized danger.

Basic economics teaches that the price of a given product or service is determined by supply and demand, without giving much consideration to those instances where the price is also influenced by taxes, laws, and habit. For example, oil is a finite commodity located only in a few specific places, generally influenced or controlled by foreign governments that find it useful for political purposes. The federal monitoring of shipping costs, protection of the shipping routes, governmental fiscal policies, and massive antipollution programs are all major factors in determining the ultimate value and availability of a product such as oil. Recognizing that this product is going to be unavailable by the end of this century, and that the economic effects will be tragic if we do not anticipate and plan for its disappearance, it makes sense to establish a federal program to start moving away from oil immediately and to adopt the oil production goals shown in the table that follows.

Since the oil crisis is not going to occur within the next couple of decades, no US government programs have yet been begun or planned to make dramatic reductions in the use of oil as a transportation energy source. Instead, we permit and encourage the production of oil for profligate vehicles such as vans and sport-utility automobiles, producing many cars with low values of miles per gallon, while other countries are producing cars capable of up to sixty miles per gallon.

We need to produce more energy-efficient vehicles, as described in section D of chapter 4. There we see some of the anticipated changes in future transportation power as motor technology makes the transition from conventional vehicles to hybrids, electric cars, fuel cells, and ultimately pure hydrogen vehicles. At the present time most automobile manufacturers are in the process of researching and developing these new technologies, although not under any government support program. It is proposed here that the need for oil in transportation should be reduced at the following rates, particularly if supported by economic and tax policies:

Year	US oil usage (in quads)
2010	40
2020	35
2030	30
2040	20
2050	10
2060	5 (for other than transportation, ultimately replacing other sources)

2. Coal

Of fossil fuels, coal is the most plentiful in the world. As described in section B of chapter 3, we can predict that at present consumption levels, the United States will run out of oil and gas in less than 100 years, whereas the coal supply will last for more than 200 years. However, we must recognize that it is dirty, polluting, and difficult to mine. A sensible energy policy for this country, while making the shift to renewable energy sources, is to develop the technology for producing liquefied and gasified forms of coal, while simultaneously cleaning it up significantly. The proposed decrease in annual coal usage not only will be beneficial for national health and the environment but also will make available much land area that is currently devoted to coal production.

3. Natural Gas

Of the various forms of fossil fuels, natural gas is the cleanest and safest energy source. It should not be confused with gasoline, which comes from petroleum and is the primary product we use to propel today's automobiles.

Natural gas is primarily composed of methane (CH_4). It is odorless and clean-burning and can be transmitted throughout the country through a network of pipelines. In the twentieth century, it began to be used to heat homes and operate appliances such as oven ranges and boilers. The current known national gas reserve in the United States is estimated to be about 1,200 quads, and this number could conceivably be doubled if all the unknown reserves were found.

4. Oil Shale and Tar Sands

Oil shale and tar sands are fossil fuels described in section B of chapter 3. In view of the proposed plan to minimize and ultimately eliminate the use of oil as a primary energy source, there is no logical reason to consider complementing this fuel source with oil shale and tar sands. Large-scale production is economically unfeasible, environmentally damaging, and politically destructive. These two fossil-fuel sources are therefore not included in future planning.

C. Plan for Postponing the Use of Nuclear Power

Nuclear power, discussed in chapter 3, section C, is initially attractive and is being used as a practical energy source in a number of applications in the United States, as well as around the world. Unfortunately, however, there are a number of significant problems that we do not yet know how to solve. Principal among these is the fact that spent fuel from nuclear power plants is toxic for an unlimited time, and we have not come up with a safe and permanent storage facility for this radioactive waste. Transporting nuclear fuel can also be dangerous, as is the possibility of an accident in nuclear power plants, such as Chernobyl, Three-Mile Island, and Fukushima Daiichi in Japan.

In view of this, it is not wise to include nuclear energy as one of the potential sources that the United States should depend on in the foreseeable decades. It therefore disappears in the table of chapter 7.

D. Plan for the Use of Certain Renewable Energy Sources

1. Hydropower

Hydropower was discussed briefly in chapter 3, section B-8, and in chapter 4, section H. Based on ocean and river flow, waves, and tides, it is a national source of clean and reliable power and electricity, as well as a source of irrigation, flood control, and recreational activities. Offsetting the advantages of hydropower as a free and reliable energy source, there are some economic and environmental concerns that need investigation. For example, dams produce reservoirs that may contain stagnant water, trapping sediments and producing algae growth. This can cause disease in the water just below the hydropower plant and turbine. In addition, a dam can block the natural movements of upstream-moving fish such as salmon and sturgeon and can kill or injure downstream-moving fish. These issues are currently being tested and evaluated in governmental research and development programs, and potential solutions involve the use of fish ladders, underwater screens, and other approaches.

On a worldwide basis, 20 percent of all electricity is generated by hydropower, whereas in the United States, it is approximately 10 percent. Since the primary form of energy captured by hydropower involves conversion to electricity, for example, through turbines, it is convenient here to express this power in the form of megawatts (MW), as defined in Appendix 3. When convenient, we can convert the megawatts to quads, using the equation 1 quad = 33,500 megawatts of power for one year.

During the last decade, many federal agencies have been involved in measuring and estimating the hydropower capacity in the United States. The results have varied greatly, ranging from the Federal Energy Regulatory Commission (FERC) estimate of 70,000 megawatts (MW) to the US Army Corps of Engineers' estimate of 580,000 MW. More recent analyses by a Department of Energy agency led to the conclusion that a practical estimate of the potential hydropower capacity in the United States is about 30,000 MW.

In 1998, the Hydroelectric Power Resources Assessment (HPRA) database did a study of all possible hydropower sites and the power potential of each and concluded that there is a gross power potential of 300,000 MW. Only 10 percent of these sites have been developed, and in most of the rest of them, development is not feasible. Most are in unlikely zones, and many would be in violation of current law. The rough conclusion is again that 30,000 MW is the power level of feasible hydropower development at the present time.

At this level (30,000 MW for a full year), a federal program undertaken in the present decade should produce a confident annual supply of one quad of national energy. It is further assumed that as new sites are developed, the total hydropower capacity will double near the end of this century.

2. Wind Energy

As described in chapter 3, wind energy is a converted form of solar energy and is used at present primarily for the production of electricity. A typical horizontal-axis wind machine—for example, a windmill with three blades and a 250-foot span–at a suitable location can produce a million kilowatt-hours (kWh) of electricity per year. Similar in benefits to hydropower, wind energy causes no pollution, generates no wastes, depletes no natural resources, and is free after the windmills are built, although they require periodic maintenance. Wind farms are power plants that may have dozens of wind machines scattered over large areas where the wind blows almost continuously. Such areas may include open plains or shorelines with average wind speeds of at least fifteen miles per hour.

Small wind turbines may be used to provide electrical power for individual residences or businesses. In such cases (e.g., a 10 kW wind turbine), the rotor diameter might be less than twenty-five feet, and the power generated in a year can be about 10,000 kWh, which corresponds to the annual electricity consumed by the average US household. On the other hand, wind turbines are now being designed with ratings of 5 MW; these can produce sufficient electricity to power 1,500 homes.

The "rating" of a wind turbine is the power that it can produce if it operates at maximum output 100 percent of the time. The average wind turbine in the western United States operates between 60 percent and 90 percent of the time, often generating significantly less than full capacity. Therefore, the actual amount of power produced in a given time period, known as the "capacity factor," is generally in the order of one-third to one-half of the wind turbine rating.

Today's wind machines are highly reliable and are rarely out of service for maintenance. However, the wind may blow steadily for a while and not at all on occasion. Therefore, even though much of the household needs can be provided by wind turbines, an energy storage system may be required, and there should generally be a tie-in to a utility company grid. The utility will provide the energy supplement when needed, and it generally will buy the excess energy when the wind is blowing and the user is receiving more electricity than is needed at the time.

In the past, all US wind projects have been on land. However, wind turbines can also be located offshore, where the wind can blow harder and more steadily. For example, Cape Cod is an attractive potential area for a US offshore wind farm. Such sites should be outside of normal shipping lanes and are not expected to have significant effects on the seabed or marine life.

Some have expressed concerns that wind machines can result in land erosion, bird kills through collisions, visual impacts, and noise. All of these objections can be dealt with. Land erosion can be prevented through proper installation and landscaping. Bird kills are low and relatively insignificant compared with collisions with buildings and airplanes. Through proper design and with so much cost benefit, the aesthetic concerns should fade away quickly. And the noise problem has been largely eliminated through careful location and improved engineering.

We can expect wind energy to supply about 20 percent of the nation's electricity. To do this, and to take advantage of the wind-rich High Plains, an electrical transmission system to cover that area and tie into the rest of the country must be designed and developed. At the present time, US electricity generation is in the order of fifteen quads per year. Today, very little of this comes from wind. However, with an active development program, and with Texas and California leading the way, it is anticipated that wind generation will provide the nation with one quad of electrical energy in the next decade and will grow to more than four quads by the end of the century. This could represent a growth of more than 100,000 jobs nationwide, with many companies throughout the country entering the wind turbine market. Most wind technology development and research is being led by the National Renewable Energy Laboratory in the US Department of Energy.

One attraction of the use of windmills is that they can provide significant additional income to owners of 100- to 200-acre farms, with no more than two acres devoted to the windmill operation. Such wind farms can ideally replace natural gas for electricity production, providing economic benefits while simultaneously reducing air pollution.

3. Geothermal Energy

The amount of geothermal energy currently being extracted and used for electricity in the United States is significantly less than one quad. On the other hand, the earth contains a vast geothermal energy resource, and it is estimated (by the Geothermal Energy Association) that the United States alone could generate more than a quad of electrical energy by the year 2020 if a practical investment was made to do so.

In many regions where the magma has penetrated the upper crust of the earth, the rock is permeable so that the water or steam circulating through it is extremely hot and can be withdrawn for practical use. Geothermal power plants fall into three primary categories: the first is dry steam, where the hot steam (over 450 degrees F) is used to directly power electric turbines; the second is flash steam, where hot water (over 360 degrees F) is pumped from the reservoir, where it vaporizes into usable steam and then is injected back for reheating and reuse; and in the third category, the hot water is extracted and used directly in a mechanical heat exchange system for use in buildings and industrial processes. Opportunities for such applications include crop treatment, horticulture, and snow and ice melting.

One aspect of geothermal energy that is receiving increasing attention is the use of geothermal heat pumps, often referred to as ground-source heat pumps (GSHP). These can be used for water heating as well as space heating and cooling, without involving any combustion or conversion to electricity. The heat pump operation is based on the fact that the temperature of the earth a few feet below the surface is constant, even though the air above the surface varies from extreme heat to subzero cold. The heat pump uses a series of pipes, known as a "loop," that carry water back and forth from the ambient air to the surrounding soil and that also may circulate, when required, into the house's hot water tank. During the winter, heat is transferred from the earth to the house, and in the summer it goes the other direction, thus providing cooling. Approximately 50,000 geothermal heat pumps are presently being installed in the United States each year.

Geothermal power is currently used in more than twenty countries around the world, with Iceland using it to produce more than 50 percent of its electricity. In the United States, major geothermal areas include California, Nevada, Oregon, and Idaho. If emphasis is put on the direct use of energy from heat pump applications, it is not unreasonable to anticipate that geothermal energy will approach ten quads of US use before year 2100.

4. Biomass Energy Sources

Much of the nation's unused agricultural land could be used to grow biomass crops to reduce greenhouse gas emissions, simultaneously generating electricity, heat, and liquid fuels. Biomass contribution to US energy is currently about 3 percent, but significant increases are possible. The amount of energy developed annually in biomass is more than five times the country's total energy consumption. The primary categories we can exploit are wood, organic wastes, and biofuels. The concept of "wood" brings up concerns about the destruction of forests, but thinning forests generally improves their health and reduces the risks of fire. Also bear in mind that more than half the energy in many developing countries in Africa and Asia comes from wood, producing some increase in carbon dioxide in the atmosphere.

Municipal solid wastes include yard trimmings, pallets, wood packaging, paper, cardboard, and plastics. In the past, much of this has gone into landfills. However, it is now being recognized that this can be an ideal source for biomass energy production.

Because corn is a plentiful product, many people assume that it is one of the sole sources of ethanol, an attractive biofuel. However, we should recognize that there are many other sources for the production of ethanol. It can be made from wheat, barley, potatoes, corn stalks, or switchgrass. One source that is often cited is animal manure converted to liquid or gas. Although interesting, the energy generated is not worth the expense. Rather, much consideration should be given to switchgrass, a perennial unlike corn and sugarcane, which is a great source of biofuel energy.

All motor vehicles currently using gasoline could operate just as efficiently if they were operated using a blend of 90 percent unleaded gasoline and 10 percent ethanol (known as E10). Also, as noted previously, many new cars are known as flexible fuel vehicles (FFVs), and they can operate satisfactorily with the percentage of ethanol going as high as 85 percent (called E85 fuel). A team of scientists sponsored by the US Department of Energy is working intensively in this area, devoted to making ethanol from switchgrass at prices competitive with fossil fuels.

Chapter 6—The Primary Solution: Direct Solar Energy

In previous chapters, we have discussed the sun and the role it plays in keeping the earth alive and functioning. It is a star that was formed more than 4 billion years ago. Today it is a burning mass of gases, with hydrogen being the most prevalent. The sun radiates energy in all directions, and less than a billionth of that radiation intercepts our planet. Most of this radiation is called the "visible spectrum." The rest is known as infrared, ultraviolet, X-radiation, gamma rays, and radio waves. Some of this radiation, such as ultraviolet (UV) and infrared (IR), is absorbed by our atmosphere and does not reach the earth. Other radiation is absorbed or reflected, as shown in Figure 1E-1.

There are periodic changes in solar radiation, evident by sunspots and bright regions, known as *faculae*, which appear periodically. In general, however, the total solar energy reaching the earth can be treated here as a constant, since the variation is essentially not significant for our purposes. We recognize that the sun plays a strong role in the changes in earth's climate. Its role does not change very much, however, as the prime source of energy to maintain humans' existence on the earth. The previous chapters have discussed the delivery of indirect solar energy, via hydropower, wind, and biomass, and the transfer and interaction of various energy forms. In this chapter we lay out a plan for civilization's practical use of direct solar energy, concentrating for the present on the United States.

The general public and many of its governmental representatives are not aware of the fact that solar energy on a massive scale is technologically available today. But there are many forces and constraints against its universal adoption; these are primarily economic, political, and sociological and fall into three principal classes of arguments. Following are the arguments and counterarguments:

Argument 1: Solar energy is too expensive to collect, transport, and use.

Counterargument: This is true, so let's change the economics. Regulation can be blatant, or it can be subtle, such as through the use of penalty and incentive taxes. If burning high-sulfur coal produces air pollution, let's add the cleanup costs to the cost of the coal. If the burning of oil eliminates its availability for other purposes, let's add a surcharge to the oil representing the cost of developing substitutes. If the use of solar energy can lead to independence from Middle East pressures, let us provide significant tax incentives to motivate the US public to demand this energy source. Thermal solar panels could become economically attractive to many new homeowners if the government were to provide a subsidy of at least a billion dollars a year. Such a federal expenditure in a single year would recover itself through savings in oil import costs in less than a decade. *Petroleum Intelligence Weekly* recently reported that such costs are now exceeding 400 billion dollars annually.

We should make the decision to use solar energy on the basis of health, security, and quality of life and then strive to adjust the economic tools to make it financially attractive. Consideration must be given to international trade impact and to our continuing relationships with present fuel-supplying nations. There must also be a national commitment and a national understanding if the required legislation is to be developed and enacted. This brings us to the second of the three arguments.

Argument 2: The application and the techniques for using solar energy are not universally understood.

Counterargument: Educate the world's people. Massive amounts of energy are wasted because of ignorance and misuse. Developing-world citizenry collect meager amounts of wood or dung to burn and then lose more than 90 percent of the energy because of bad design of their earthen ovens. Fast-growing plants, biomass conversion, and conservation techniques must be developed and disseminated. People in general must be made aware of all the forms of solar energy that they can use and the fact that much of this energy is technologically available to them today. Such an education program is possible, but it requires the dedication and concerted efforts of our nations' leaders.

Argument 3: The disruptive effects of a sudden switch from oil would destroy the multinational oil companies and produce worldwide economic chaos. Therefore, the big oil companies are not going to permit this to happen.

Counterargument: There is no need for the creation of new corporate giants. Let us give the solar energy responsibility to the oil companies. The industrial structures are in place for developing, servicing, and distributing energy to the world. We need the know-how, management ability, and cooperation of such organizations if we are to undertake a universal program to exploit solar energy. It makes no sense to destroy existing corporations and create new ones that must be just as large and just as powerful. With proper regulation and profit motivation, the oil-distributing multinationals can make the switch to solar energy with the leadership talent we need and with the least economic disruption.

The intent here has been to develop the argument that the energy "crisis" is not one of unavailability, but rather one of economic imbalance. As far as energy sources are concerned, we have and will continue to have all the energy we need. In making the decisions regarding the development, distribution, and use of these sources, it is time to make judgments and take actions that are based on concern for the future, compassion for people, and hope for the common good of society. We should disregard the comparison of prices that are artificially constrained by man's inherent selfishness and shortsightedness, by vested interests unwilling to change, and by oil-distributing multinationals.

Concentrating on the United States, we now examine some scenarios for using direct solar energy in a way that will best suit society's needs, now and for the future.

A. Solar Thermal

As long as the sun shines on our rooftops for an average of about six hours per day, it would make sense to utilize the energy for changing the temperatures in our homes and offices. More than 25 percent of the total energy used in the United States is for air and water heating.

As described earlier, solar thermal collectors are flat plates used to heat swimming pools, water for residential and commercial use, and interior space. These panels are generally insulated boxes that have a glass cover and a dark absorber plate. In many of these panels, water flows through tubes in the plate until it is hot, and it is then moved into an insulated storage tank. In the last thirty years, these systems have become increasingly sophisticated, using pumps, non-freezing heating fluid, and heat exchangers.

In the simplest systems, the water heated directly by the sun is the same water stored in the hot water tank and circulated through the building. Since it is susceptible to freezing at times in some locations, the design may require some additional safeguards. A more complex system, known as "closed loop," uses an antifreeze solution to receive the sun's energy and travel through the building, without ever mixing with the building's water supply. In both cases, the collectors look like skylights, and they are mounted on panels on unshaded areas of rooftops facing south, southwest, or southeast. Over 20 million square feet of solar thermal collectors are being produced in the United States each year. A solar thermal system reduces gas or electricity costs significantly, usually repaying the initial investment in less than five years.

If a national program were undertaken to incorporate solar heating, cooling, and ventilation in more than half of the existing commercial and residential buildings, we could expect that in a short time, the savings in energy would be in the order of three quads, as shown in section B-3(a) of chapter 3. If, in addition, all new structures were to require an engineering evaluation of geographic location, configuration, orientation, and air-conditioning systems, the operating costs and building life cycle could be improved significantly.

B. Solar Voltaic

As we consider the use of energy for various heating and industrial purposes, we find electricity to be a convenient carrier of energy, even though it is not the original source. To create this electricity, we have been referring regularly to the use of turbines and generators, but as described in chapter 3, electricity can be produced directly from sunlight by using photovoltaic (PV) cells.

When more than a single cell is required, they can be electrically connected behind a glass sheet and are then known as solar PV panels. Just as in the case of the previously described thermal panels, these panels are arranged in arrays that can be mounted to a flat roof, but they now are producing electricity instead of heat. In the future we can expect to see many rooftops containing both solar thermal and solar voltaic panels.

Some of the advantages of PV panels are that they are pollution-free and require little maintenance. They can be tied into local utilities through grid connections, so that they partner with the utilities in providing electricity to the buildings. In fact, at times when they are providing more electricity than the building requires, the excess can be transferred to the utility grid, with the cost being credited to the building owner.

A recent technological breakthrough has been the development of a new "multi-junction" solar cell that achieves an energy-conversion efficiency of more than 40 percent, which produces systems in which the energy savings within the first three years are equivalent to the total energy cost of producing and installing the system. After the first three years, the energy is free and completely nonpolluting.

Currently, the PV capacity in the world is about 6,000 MW, including the US capacity of 90 MW. The applications are growing fast. In addition to roofs and walls, the panels can be installed some distances from the target buildings, using batteries and cables to store and then transmit the electricity.

In section B-3 of chapter 3, an estimate was made of the total US energy that could be obtained from roof-mounted thermal panels, and the conclusion was that within a decade, we could plan on a total of about three quads. If we now assume that the same homeowners will install an equivalent area of PV panels on the same roofs, but recognize that photovoltaic systems are less efficient than standard thermal panels, we can predict that the total national energy use of photovoltaics will probably be in the order of one quad.

C. Concentrated Solar Power (CSP)

There are a number of significant considerations associated with the use of land for concentrated solar power (CSP) applications. The primary one is the realization of the amount of land available for CSP use. Another is the recognition that at the present time, vast areas of land are being devoted to nonrenewable energy sources, such as coal mining and excavations, and millions of acres for oil and gas development. These land areas can all be converted to other purposes when the country becomes a user of solely renewable energy sources. Other decisions must be made regarding the type of CSP to select, such as troughs, dishes, or towers. The values of available energy discussed in this section are approximately the same for any of these CSP types.

We have shown previously (Table 3B-7) that every square yard of US land exposed to direct sunlight receives an average of 833 watts (0.833 kW) of power. If the sun's light provides an average of six productive hours of energy per day, then a single square yard receives about 5 kilowatt-hours (kWh) of solar energy in a single day. The CSP systems discussed in this section generally can achieve an efficiency of 30 percent, so that the useful energy production is in the order of 1.5 kWh per square yard per day. This is equivalent to 7,260 kWh per acre per day, or 4,646,000 kWh per square mile per day. As seen in Appendix 3, one quad of energy is equal to $293(10)^9$ kWh, so one square mile can produce $15.86 (10)^{-6}$ quads per day $(4,646,000/293[10]^9)$ and 0.00579 quads per year. If we establish as a goal that CSP should provide 70 percent of the nation's energy use fifty years from now (which is in the order of 107 quads, as shown in Table 7-1), we can determine that the required solar-collecting land area in square miles is less than 19,000 square miles (107/.00579).

The United States has a total land area of 3,600,000 square miles. If we use 19,000 of these for the capture and use of solar energy, this will represent about one-half of 1 percent of the country's land area, and far less than the land devoted to forests, pastures, crops, or special uses.

Before concluding that the plan described here is overwhelming, we must recognize that the proposed actions in this book will solve the nation's energy problems! In the century ahead, we will eliminate the use of fossil fuels, lose all dependency on foreign suppliers, and provide all these benefits at a reduced total cost (over the next five decades and forever thereafter). We need to recognize at this time that the initial installation and adoption of CSP solar energy systems will have an obvious effect on the national landscape, but it will make significant beneficial changes in our lifestyle and political and economic comfort throughout the next century.

This book is not proposing a specific layout for how and where to locate the solar power plants to utilize 17,000 square miles of area for capturing solar energy.

There certainly will be many solar power plants distributed throughout the United States, and in all probability, the largest of them will be located in the West and Midwest. The concentrated energy will generally be used to produce heat, which may be stored or directly converted to electricity when desired. An optimum selection of locations and sizes of power plants will require the input of engineers, local governments, and economists. Significant land is available, and the know-how exists to build the power plants that will provide the country with all of its independent energy needs in the foreseeable future. This is shown in chapter 7.

D. Solar Energy Storage Considerations

In the application of solar thermal collectors or photoelectric panels, the collected heat and electricity are used directly. However, in the use of CSP systems, consideration must be given to the energy forms that are collected, as well as the questions of how to store and how to utilize this energy. The direct energy is captured from the sun's rays, and we must determine how to use it when the sun is not shining or the sky is overcast.

Among the various parameters that may be considered are the use of common or liquid batteries for electricity storage, insulated high-pressure water tanks, and turbines for power production. We can also take advantage of recent MIT techniques to use the sun's energy to split water into hydrogen and oxygen gases and later recombine them in a fuel cell when we want the electricity. Also, pumped-storage hydroelectricity may be used by pumping water into high-elevation reservoirs, for use as a power generator when the water is released. One example of this option for many years has been Niagara Falls.

E. Transition to a Hydrogen Economy

Reference is made to hydrogen and its applications in several sections of chapter 4, as well as to its potential use in fuel cells and automobiles of the future. If we combine the wide potential of hydrogen applications with the surplus availability of solar energy, we can envision a social and technological environment in which world energy needs will be met easily and economically. First, we must recognize that hydrogen is *not* a natural fuel. It is a means of storing or transporting energy after it has been separated from its associated elements, such as carbon, nitrogen, and oxygen. If this separation can take place without the creation of polluting gases (e.g., carbon monoxide, carbon dioxide, etc.), the pure hydrogen can become an easier and cheaper carrier of energy than electricity or fossil fuel sources. Society's challenge, therefore, is to store, transport, and use hydrogen that has been separated and packaged using only zero-emission generation techniques. Basically, these techniques will not continue to use the current dominant process of reforming, which produces hydrogen from hydrocarbons, simultaneously polluting the atmosphere.

Researchers today are investigating the use of titanium ceramic photoelectrodes to harvest the sunlight and split water to produce pure hydrogen, with no moving parts and no emitted pollutants. Scientists from the University of New South Wales believe that all of Australia's energy needs could be provided in less than a decade, by using rooftop panels on 1.6 million houses. Titanium dioxide happens to be plentiful and cheap.

As we look toward the future, it is very apparent that

- the world's requirement and use of energy will continue to grow;

- fossil-fuel energy sources are limited, are being used up rapidly, and will generally be unavailable in another century;

- biomass sources are available but are extremely limited when compared to world needs;

- wind, hydropower, and geothermal sources are attractive and should be used, but again represent a small percentage of world energy requirements;

- nuclear energy sources can continue to be developed but carry the high-probability risk in development and storage of waste; and

- the available solar energy that can be captured safely and economically is a significant portion of the total annual solar energy received by the earth, which is 10,000 times the world energy used by humankind.

An interim source is therefore required to capture the necessary solar energy, store it, transport it to desirable locations, and transfer it to the forms that are useful to society. This interim source is generally recognized to be hydrogen, which is briefly described in several sections of chapter 4.

Earlier in this book, we discussed electricity generation through the use of excess solar energy. One attractive possibility is the coproduction of electricity with hydrogen. The hydrogen can then be used to fuel a vehicle, serve as an unused energy storage system, or supply reliable electricity to off-grid communities.

At present, a number of scientific organizations around the world are concentrating their efforts on developing low-cost photoelectrolytic technology to use the sun to split water into hydrogen and oxygen with no pollutants generated. Such activities are going on in many places in the United States, as well as Australia, Israel, Japan, Germany, England, and elsewhere.

Chapter 7—A Rational Energy Plan for the Twenty-First Century

This book is dedicated to the creation of a sensible national energy program to provide our society with permanent, clean, and low-cost energy to satisfy all of our needs now and into the future. The actions to accomplish these objectives will require governmental and public support, and so the quantities and uses of various types of energy have referred principally to the United States, but the basic concepts are similarly applicable to all the countries of the world, assuming that their governments and populations are rational and dedicated to the interests of their people.

With an emphasis on US interests, a national energy plan should include three primary considerations: (1) use of energy sources that are not dependent on foreign control (such as imported oil); (2) government support and marketing of the development of solar systems; and (3) a tax structure that suitably supports and/or rejects the various actions that can be taken.

These considerations are the basis of the rational energy plan developed throughout this book and summarized in tabular form at the end of this chapter. To put this into action, the president and Congress must take leadership roles through speakers, declarations, legislative acts, and citizen education. If proper steps are taken to achieve the numbers in Table 7-1, the United States will be independent and economically strengthened.

Table 7-1 presents the principal conclusions of this book in a simple and logical way, and it is recommended that the reader give careful consideration to the data presented herein. The following key points and comments refer to the various items in the table and (hopefully) provide the reader with justification that the proposed energy plan is sensible, simple, and in our individual best interests.

Item 1: Recognizing that most of this book has been written in the year 2010, the goal here is to take logical steps over the course of this century to adopt an ideal energy philosophy. The target dates are shown for the first three decades and then for twenty-year intervals until the year 2100.

Item 2: In predicting the total US quads, it is recognized that our population is currently growing at a rate of about 1 percent per year, much of which is due to immigration. We have also assumed that energy use per US citizen will continue to increase, but will then level off as we develop continuous increases in efficiency.

Items 3, 4, and 5: Programs should be undertaken immediately to start reducing the use of oil, coal, and natural gas, with specific plans showing how these will fade out significantly over the next hundred years. At the end of chapter 3, it is shown that current US use of fossil fuels is about eighty-nine quads per year. However, this will have to go down as costs go up rapidly and world resources start to disappear.

As noted in chapter 5, two fossil fuels often recognized are oil shale and tar sands, but these can be environmentally damaging and economically infeasible, and their net contributions to world energy may well be negative. Therefore, they are not included in the proposed energy plan.

Item 6: Nuclear energy has many promises, but there are also several unknowns. A primary problem is the disposition of radioactive waste; a second is the danger associated with a nuclear plant failure, such as those in Chernobyl, Three-Mile Island, and Fukushima; and a third is the fear of the use of nuclear bombs for terrorist activities. For such reasons, the proposed plan terminates the use of nuclear energy for the present.

Items 7, 8, and 9: Renewables such as hydropower, wind, and geothermal are recommended for development and exploitation, although their contributions to total US energy are relatively small. (They can play a much larger role in many small countries in Europe and Asia.)

Item 10: As described briefly in chapters 3 and 5, biomass has significant growth potential, with particular emphasis on switchgrass as an attractive source of biofuel (ethanol). It is anticipated that by the end of the century, this will be a larger contributor to the nation's energy than any of the previously listed standard sources.

Items 11 and 12: Two forms of solar energy that are currently being developed in many countries involve the installation and utilization of (roof-mounted panels that produce limited heat (solar thermal panels) and/or limited electricity (solar voltaic panels).

It is logical to recognize that conversion of solar radiation to low-voltage applications should be exploited, particularly in areas that are remote and not parts of any grid.

Item 13: And now we come to a dramatic conclusion: If all the energy sources in lines 3 to 12 are added together and subtracted from the total US quads required in line 2, we arrive at line 13, showing that the annual requirement for concentrated solar energy will reach a value of 107 quads in the year 2100. As described in section B-3(b) of chapter 3 and section C of chapter 6, this energy can be provided by intercepting and using the solar energy that falls on one-half of 1 percent of US land area.

So there is no energy crisis. What we must do is devote a major national effort to the capture and use of the massive amount of solar energy that presently is intercepted by the earth and immediately radiated away. Further consideration should be given, however, to the two key issues in chapter 5, each making use of hydrogen as a major carrier of energy. In the first case, we should be modifying our transportation technology, with the ultimate development of vehicles dependent on hydrogen rather than gasoline, batteries, or combinations such as hybrids. In the other applications, we will go directly from solar to hydrogen and then use the hydrogen as a plentiful and clean energy source.

It is time for us to recognize that society must prepare itself to use energy sources other than fossil fuels. Fortunately, this need not be accomplished in a single instant, or even in a decade. But we should start planning now to develop the alternate sources and to switch over to them in a reasonable time period. Simultaneously, we can start to decrease our usage of fossil fuels, letting them fade away in a time period that is reasonable and not critical; these goals are illustrated in the table that follows. The world will then be able to continue to develop positively, without the energy concerns we express today.

Table 7-1
US Energy Goals Expressed in Quads

1	Year	2020	2030	2040	2060	2080	2100
2	US total quads	110	118	126	140	150	154
	Energy source						
3	Oil	35	30	20	5	3	1
4	Coal	23	22	20	15	8	4
5	Natural gas	23	22	20	17	14	9
6	Nuclear	3	0	0	0	0	0
7	Hydropower	1	1	1	1	2	2
8	Wind	1	2	2	3	4	5
9	Geothermal	1	1	2	3	4	5
10	Biomass	5	7	9	10	12	12
11	Solar thermal	2	3	4	5	6	7
12	Solar voltaic	1	1	1	1	2	2
13	Concentrated solar	15	29	47	80	95	107

The reader should recognize here that the table relates only to the United States, showing how 100 quads of energy usage at present will be able to grow to 154 quads by the end of the century. During the same time period, the world usage of 450 quads is also expected to grow appreciably. However, the recommended policies described here are recommended goals for US action, leaving it to the rest of the world to take actions that will solve their energy problems. As shown by Table 7-1, combined with the information presented in this book, the "energy problem" can easily cease to exist, through utilization of energy sources that are available and economical and particularly through taking sensible advantage of the massive amounts of energy offered by the sun.

Chapter 8—A Brief Reminder of Economic Considerations

The objectives of the first four chapters were devoted principally to defining "energy" and describing its forms, sources, uses, and availability. Chapters 5 to 7 attempted to portray a realistic description of each of these sources, showing how long they can last and what quantities are available and usable. This information is summarized in Table 7-1, which shows that the availability of solar energy will satisfy society's needs far into the future.

A matter that may be of concern to many readers is that there has been no emphasis on or detailed description of the costs involved in taking the recommended actions to achieve the goals of Table 7-1. The response to such a concern is that a detailed economic evaluation may produce some changes in the numbers in the table, but these changes would not be very significant when compared to the major conclusion that the table and data present—*during the twenty-first century, solar energy in the United States will climb from 3 percent today to more than 75 percent in the year 2100.*

Following are some examples of the economic considerations that may affect the values in Table 7-1:

- The use of fossil fuels shows dramatic decreases. This occurs primarily because the fuels are not renewable, and the sources will eventually disappear. As this happens, the costs per unit will climb significantly, until they are far more costly than the other sources.

- At the same time that coal availability goes down, the cost of handling and transportation will climb.

- The use of natural gas should take into account the location and availability of pipelines.

- In considering biomass, it is important to pick source locations that minimize transport costs. For example, it is desirable for the plant to be within fifty miles of the source material.

- A 2002 study showed a cost of $60 million in one year to build a 40-million-gallon-per-year ethanol plant. It is also of interest to note that biomass producers are usually farmers who have 25 percent to 40 percent equity.

- When using CSP systems in non-developed areas such as deserts, other problems can arise, such as the unavailability of water. This must be considered in financial analyses of such systems.

- Taxes and incentives should be designed so that capital investment costs are annualized to be no more than present costs for a reasonable number of years and should then drop to zero. For example, consider a situation in which a property owner pays $10,000 a year to purchase electricity, and it is determined that $75,000 will be the total cost for installing a CSP system to produce all the solar energy required to maintain the property. In such a case, it should be made legally possible to pay off the investment in less than eight years, after which all subsequent energy usage is essentially free forever, except for periodic (small) maintenance costs.

Appendix 1—Some Brief Definitions for the Non-Scientist

Matter: A material that occupies some space.

Mass: A quantity of matter forming a body of indefinite shape and size.

Element: A substance that has unique properties and cannot be separated into different substances by ordinary chemical methods. At the present time, 116 elements have been identified, of which 93 are known to occur naturally on earth. Every element has a different number of electrons that orbit around the atomic nucleus, where a similar number of protons are located. Examples of common elements are oxygen, hydrogen, nitrogen, and carbon.

In chemistry texts, elements are usually presented in a periodic table, which groups the elements in a way that displays their various properties. Each element can also be identified by its name, its symbol, or its atomic number (see below).

Compound: A substance composed of two or more elements in chemical combination. For example, water is a compound made up of the elements hydrogen and oxygen.

Atom: The smallest unit of a chemical that which has the properties of that element (refer to Figure 3C-1). An atom consists of a central core (the nucleus) and negatively charged electrons that orbit the nucleus.

The nucleus consists of neutrons (uncharged particles) and protons (positively charged).

Atomic number: The number of protons in the nucleus (when uncharged, this is balanced out by the same number of orbiting electrons).

Atomic mass: This is equal to the total numbers of protons and neutrons in the nucleus.

Isotopes: Atoms of the same element (same number of protons and electrons) but with different numbers of neutrons in the nucleus (so their atomic masses are different, even though the atomic number is the same).

Molecule: The smallest particle of a compound that has all the chemical properties of that compound. For example, one molecule of water is made up of two atoms of hydrogen and one atom of oxygen. The symbol for the water molecule is H_2O.

Some of the energy forms described in chapter 1 are based on the kinetic energy of the particles defined in this appendix. For example, thermal energy is dependent on the motion of individual molecules; chemical energy is stored in the chemical bonds between atoms; electrical energy comes from the interactions between charged particles; and nuclear energy is derived from the potential stored between the constituents of the individual nuclei.

Appendix 2—Some Basic Elements of Energy Capture in Living Matter

Hydrocarbon: Any organic compound composed solely of the elements hydrogen and carbon. Various hydrocarbons differ in their proportion of hydrogen to carbon. Many hydrocarbons can burn in air, using oxygen to enable them to emit energy, producing waste products such as carbon dioxide (carbon and oxygen) and water (hydrogen and oxygen).

Carbohydrates: A large class of chemical compounds that are produced by green plants from carbon dioxide and water (see entry for photosynthesis). Carbohydrates include sugars, fats, starches, and cellulose and are important food elements, supplying proteins and nutrition to plant and animal life.

Chlorophyll: A green pigment within the cells of plants that absorbs a significant portion of the visible energy from the sun.

Photosynthesis: The process in which green plants use the energy of sunlight to manufacture carbohydrates from carbon dioxide and water in the presence of chlorophyll, which is concentrated in the leaves. Photosynthesis provides the source of energy that drives all the metabolic processes of plants and animals.

Carbon cycle: Carbon is a very widely distributed element throughout the world, being part of more compounds than any other element. The study of carbon compounds is called organic chemistry. All living organisms contain carbon, and these carbon atoms have been parts of other molecules and reused continuously throughout the history of the world. In the carbon cycle, plants absorb carbon dioxide from the atmosphere and combine it with water they get from the soil. The process of photosynthesis enables these compounds to absorb solar energy and produce foods and carbohydrates as needed. These energy-laden compounds may undergo many transformations, and when they eventually decay completely, they separate back into their original forms, and the energy is ultimately radiated back away from the earth.

Note: Photosynthesis is often known as primary production, and the release of carbon dioxide is known as respiration.

Appendix 3—Units, Measurements, and Conversions

Engineers, analysts, scientists, and laymen use many different terms and expressions to identify common units, such as length, mass, force, time, and so on. The number of these expressions increases significantly when we get into physics, mathematics, electricity, power, and energy. In the text of this book, an effort has been made to minimize the terms used and to keep them simple. However, it may be of value to the reader to see a listing of some of these terms and to have an understanding of their various relationships and magnitudes. Hence, this appendix is provided, for general education in some cases and for clarification of reader concerns in others. It includes many simple relationships and definitions for common measurements that the public frequently encounters, as well as some of the more esoteric expressions that may clarify some of the data presented in the body of the text. These definitions and expressions are grouped into the following classifications:

A. Prefixes for basic metric units
B. Expressions of distance and area
C. Expressions of mass and weight
D. Expressions of liquid measure
E. Expressions for force
F. Definitions of energy units
G. Expressions for electrical power
H. Some energy abbreviations
I. Some energy conversion factors
J. Energy contents of common fuels

The first portions (A through E) of this appendix provide simple relationships and definitions for common measurements that the public frequently encounters. The later portions (F through K) refer specifically to expressions for energy, identifying the units that are used and the relationships between these units.

A. Prefixes for basic metric units

Name	Value (in powers of 10)	Number of zeros	Prefix	Symbol
Septillion	10^{24}	24	Yotta	Y
Sextillion	10^{21}	21	Zetta	Z
Quintillion	10^{18}	18	Exa	E
Quadrillion	10^{15}	15	Peta	P
Trillion	10^{12}	12	Tera	T
Billion	10^{9}	9	Giga	G
Million	10^{6}	6	Mega	M
Thousand	10^{3}	3	Kilo	k
Tenth	10^{-1}		Deci	d
Hundredth	10^{-2}		Centi	c
Thousandth	10^{-3}		Milli	m
Millionth	10^{-6}		Micro	μ
Billionth	10^{-9}		Nano	n

B. Expressions of distance and area
1 yard is the basic English unit
1 meter is the basic metric unit (approximately 39 1/3 inches)

1 inch = 1/36 yard
1 foot = 1/3 yard = 12 inches

1 rod = 5 1/2 yards
1 furlong = 220 yards
1 mile = 1,760 yards = 5,280 feet

Land areas
    1 acre = 4,840 square yards
    1 square mile = 640 acres
    1 square meter = 1.196 square yards = 10.76 square feet

C. Expressions of mass and weight
    1 pound is the basic English unit = 454 grams = .454 kilograms
    1 gram is the basic metric unit of mass = .035 ounces

    1 ounce = 1/16 pound = 28.35 grams
    1 kilogram = 2.205 pounds
    1 dram = 1/16 ounce
    1 short ton = 2,000 pounds
    1 long ton = 2,240 pounds

D. Expressions of liquid measure
    1 gallon is the basic English unit = 3.785 liters
    1 liter is the basic metric unit =1.057 quarts

    1 quart = 1/4 gallon
    1 pint = 1/2 quart = 1/8 gallon
    1 fluid ounce = 1/16 pint
    1 barrel = 42 gallons = 159 liters

E. Expressions for force
    1 pound = a unit of force that gives a free mass of 1 pound an acceleration of 1 foot per second per second.
    1 newton = a unit of force that gives a free mass of 1 kilogram an acceleration of 1 meter per second per second.
    1 dyne = a unit of force that gives a free mass of 1 gram an acceleration of 1 centimeter per second per second.

F. Definitions of energy units

1 BTU (British thermal unit) is the amount of heat energy required to raise the temperature of 1 pound of water by 1 degree Fahrenheit.

1 calorie is the amount of heat energy required to raise the temperature of 1 gram of water by 1 degree Centigrade.

1 Calorie (in food ratings) = 1 kilocalorie = 1,000 calories

1 erg is the energy expended by 1 dyne acting through 1 centimeter.

1 foot-pound is the energy expended by the force of 1 pound acting through 1 foot.

1 joule is the energy expended by the force of 1 newton acting through 1 meter. It is also the amount of energy required to provide 1 watt of electrical power for 1 second (also known as the watt-second).

1 therm = 100,000 BTU

1 quad = 1 quadrillion BTU (10^{15} BTU) (also see sections I and J)

1 kilowatt-hour is the amount of electric energy dissipated in an electric circuit carrying 1 kilowatt of power for 1 hour (the unit used on utility bills).

G. Expressions for electrical power

Note that power (e.g., current × voltage) is defined as energy dissipated over a unit of time.

1 watt is the power from a current of 1 ampere flowing through 1 volt, also equal to 1 joule per second.

1 kilowatt = 1,000 watts

1 megawatt = 1,000,000 watts

1 joule (defined previously in terms of force and distance) is also the amount of electrical energy used when 1 watt of power travels through a circuit for 1 second.

1 BTU/second = 1.055 kilowatts

1 BTU/hour = .293 watts

1 horsepower = 745.7 watts

H. Some energy abbreviations

1 MJ = 1 million joules

1 GJ = 1 billion joules

1 TJ = 1 trillion joules

1 PJ = 1 quadrillion joules

1 EJ = 1 quintillion joules

I. Some energy conversion factors
1 calorie = 4.185 joules (or 1 joule = .2388 calories)
1 foot-pound = 1.3558 joules
1 BTU = 252.2 calories = 1055 joules
1 kilowatt-hour = 3,412 BTU = 3,600,000 joules = 3.6 MJ
1 quad = 1.055 EJ (or 1.055 quintillion joules) = 251.9 trillion Calories
= 293 billion kilowatt-hrs = 33,450 megawatt-years (or 1 trillion
kilowatt-hours = 3.412 quads)

J. Energy contents of common fuels
1 ton of coal = 26 million BTU
1 barrel of gasoline, oil, or ethanol (42 gallons) = 5.7 million BTU
1 cubic foot of natural gas = 1,030 BTU
1 cord of wood = 20 million BTU
1 bushel of corn = 350,000 BTU

Expressed in terms of quads:

1 quad = 38.5 million tons of coal (100 million tons = 2.6 quads)
1 quad = 172 million barrels of oil (1 billion barrels = 5.8 quads)
1 quad = 971 billion cubic feet of natural gas (1 trillion cu. ft. = 1.03
quads)
1 quad = 7.5 billion liters of petroleum
1 quad = 60 million dry tons of biomass (or agricultural residue)

Appendix 4—The Process for Producing Ethanol

1. We start with the collection of starch, a white carbohydrate substance produced by photosynthesis (Appendix 2) and found in many grains. Starch is a major nutrient in the human diet, and two of its chief food sources are potatoes and corn. In the United States, the main feedstock for ethanol is corn, in addition to grain sorghum, wheat, barley, sugar beets, cheese whey, and potatoes. Brazil has a large industry based on sugarcane (enabling the replacement of 40 percent of the country's demand for gasoline). The starch, generally in the form of kernels, is ground in a hammer mill to produce a fine powder.

2. The starch is then cooked briefly, mixed with water and enzymes, and permitted to germinate into sugars. Sugar is one of the world's most important foods, furnishing humans with most of the energy that is transferred from the sun to living plants. As defined in Appendix 2 (under "photosynthesis"), the green leaves from the plants combine the sun's energy with water from the soil and carbon dioxide from the air to form sugar and oxygen. In addition to the common sugar in our sugar bowls (sucrose, from sugarcane and sugar beets), there are simple sugars in fruits, vegetables, nuts, and milk. The first part of the sugar-manufacturing process involves washing, squeezing, cutting, and crushing these products to produce juices, as well as the by-products that can be used as food for livestock. The juices are then treated chemically, refined, heated, and subjected to evaporation, leading to a final solution of crystallized products.

3. The sugars then undergo a fermenting process called ethylene hydration, in which certain species of yeast help to produce an alcohol-water mixture.

4. The mixture is further separated by distillation to produce pure fuel ethanol.

References

Blankenship, Robert E. *Molecular Mechanisms of Photosynthesis.* Oxford, England: Blackwell Science, 2002.

Boyle, Godfrey. *Renewable Energy—Power for a Sustainable Future.* Oxford University Press, England, 2004. ISBN 13: 978-0-19-926178-9.

Butti, Ken, and John Perlin. *A Golden Thread.* Van Nostrand Reinhold Company, New York 1981.

Caldicott, Helen. *Nuclear Power Is Not the Answer.* New York, NY: New Press, 2007.

Carr, Donald E. *Energy and the Earth Machine.* New York, NY: Norton, 1976.

Chiras, Dan. *The Homeowner's Guide to Renewable Energy.* New Society Publishers, Canada, 2006.

Cohen, Stephanie. "Energy Dreams and Energy Realities." *New Atlantis,* (Spring 2004 Issue, Pages 3 to 17).

Davis, Stacy C. *Transportation Energy Data Book.* Edition 29, Oak Ridge National Laboratory, National Transportation Research Center, Knoxville, TENN., July, 2010.

Fujishima, A., and K. Honda. "Electrochemical Photolysis of Water at a Semiconductor Electrode." *Nature* 238. (International weekly Journal of Science, 1972

Goodstein, David. *Out of Gas: The End of the Age of Oil.* New York, NY: Norton, 2004.

Gratzel, Michael. *Energy Resources through Photochemistry and Catalysis.* New York, NY: Academic Press, 1983.

Green, B. D., and R. G. Nix. *Geothermal—The Energy under Our Feet.* NREL/TP-840-40665. National Renewable Energy Laboratory, Golden, Colorado, 2006.

Halacy, Daniel. *The Coming of Age of Solar Energy.* Harper and Row, New York, 1973.

Martin, Christopher, and D. Y. Goswami. *Solar Energy Pocket Reference Books,* International Solar Energy Society, Freiburg, Germany, 2010.

Romm, Joseph. *The Hype about Hydrogen.* Washington, D.C.: Island Press, 2004.

Russell, P. L. *Oil Shales of the World: Their Origin, Occurrence, and Exploitation.* New York, NY: Pergamon Press, 1990.

Schaeffer, John. *Solar Living Source Book.* Gaiam Real Goods, New Society Publishers Ltd., Gabriola Island, BC, Canada 2007.

Schwartz, Marie Sokol. "Harvesting the Sun's Energy." *Popular Science,* Design III Printing (January 1976).

Smil, Vaclav. *Energy at the Crossroads.* MIT Press, 55 Hayward St, Cambridge MA, 2003.

Sperling, Daniel, and James Cannon. *The Hydrogen Energy Transition.* Amsterdam, the Netherlands: Academic Press, 2004.

Tester, Jefferson W. *Sustainable Energy: Choosing among Options.* MIT Press, Cambridge, MA, 2005.

Traister, Robert J. *All about Electric and Hybrid Cars.* Tab Books, ISBN 0-8306-2097-4, 1982.

Walsh, Marie. "Biomass Feedstock Availability in the U.S." *Harvard Business Review, Delaware* (April 1999).

Yergin, Daniel. *The Prize.* New York, NY: Simon and Schuster, 1991.

Yount, Lisa. *Energy Supply.* New York, NY: Facts on File, 2005.

Some Prime Energy Organizations and Authorities

National Renewable Energy Laboratory (US Department of Energy)
Energy Information Administration (US Department of Energy)
US Department of Commerce
Solar Energy Industries Association

CPSIA information can be obtained at www.ICGtesting.com
Printed in the USA
LVOW060018300911

248466LV00001B/79/P